Contents

Humberto Maturana: Reflections on Bringing Forth Worlds
Pille Bunnell and Frederick Steier (Guest Editors)

Foreword
    Frederick Steier and Pille Bunnell 5

Articles

Maturana, Art and Cybernetics
    Frank Galuszka 15
Upsetting Apple Carts
    Jay S. Efran 25
Reflections on Core Ideas of Humberto Maturana in Relation to The World Cafe
    Juanita Brown and Amy Lenzo 35
Designing for Emergence: Creating Living Networks of Conversaton Grounded in Love
    Flavio Mesquita da Silva 45
The Love That Was Not Recommended: Maturana's Biology of Love
    Seiichi Imoto 55
Maturana's Path of Objectivity-in-Parenthesis
    Raul Espejo 63
Humberto Maturana: Using His Biological System in the Social Domain
    Hugh Gash 77
How My Understanding of Languaging in Non-Speaking People with Autism Has Been Informed by Conversations With Humberto Maturana
    Kathleen Forsythe 87
Consciousness: An Unresolved Question, and What Maturana Has to Say About It
    Pier Luigi Luisi 99
Humberto Maturana on Time: Zero-Time Cybernetics
    Jude Lombardi and Larry Richards 107
Animals and Humans Alike
    Lloyd Fell 123
Power Arises Through Obedience: A Conversation About the Years of Dictatorship in Chile, the Helplessness of Power, and the Freedom of the Individual
    Humberto Maturana and Bernhard Poerksen 125

Regular Features

Guest Column

Second Order Cybernetics and the End and Beginning of Philosophy
    Bernard Scott 139

ASC Column

Mutual Arisings: Conversations With Humberto
    Ray Ison 147

Book Review

Leydesdorff's Compass
    Mark William Johnson 153

The Artist for this issue is Pille Bunnell. Full color art at www.chkjournal.com.

Cover Art
Bunnell, Pille. (2021). *Luminous Consequence of Autopoiesis*. Photograph.

CYBERNETICS & HUMAN KNOWING
A Journal of Second-Order Cybernetics, Autopoiesis & Cyber-Semiotics
ISSN: 0907-0877

Cybernetics and Human Knowing is a quarterly international multi- and trans-disciplinary journal focusing on second-order cybernetics and cybersemiotic approaches.

The journal is devoted to the new understandings of the self-organizing processes of information in human knowing that have arisen through the cybernetics of cybernetics, or second order cybernetics its relation and relevance to other interdisciplinary approaches such as C.S. Peirce's semiotics. This new development within the area of knowledge-directed processes is a non- disciplinary approach. Through the concept of self-reference it explores: cognition, communication and languaging in all of its manifestations; our understanding of organization and information in human, artificial and natural systems; and our understanding of understanding within the natural and social sciences, humanities, information and library science, and in social practices like design, education, organization, teaching, therapy, art, management and politics. Because of the interdisciplinary character articles are written in such a way that people from other domains can understand them. Articles from practitioners will be accepted in a special section. All articles are peer-reviewed.

Subscription Information

Price: Individual £80; Institutional: £188+VAT (online); £233 (online & print). 50% discount on full set of back volumes. Payment by cheque in £UK (pay Imprint Academic) to PO Box 200, Exeter EX5 5HY, UK; Visa/Mastercard/Amex
email: sandra@imprint.co.uk

Editor: Jeanette Bopry, Instructional Sciences, Ret.
jeanette.bopry@gmail.com

Managing Editor: Carlos Vidales, University of Guadalajara, Mexico. morocoi@yahoo.com

Associate Editor: Sara Cannizzaro, Research Fellow, Centre for Computing and Social Responsibility (CCSR), De Montfort University, Leicester, UK. sblissa@gmail.com

Joint Art and Website Editor: Claudia Jacques, Knowledge Arts Studio, New York.
claudiajacquesmc@gmail.com; cj@claudiajacques.com

Book Review Editor: Bill Seaman, Duke University.
bill.seaman@duke.edu

Special Topics Editor: Dirk Baecker, Witten/Herdecke University, Germany. dirk.baecker@uni-wh.de

Columnist: Lou Kauffman, University of Illinois–Chicago.
kauffman@uic.edu

ASC Liaison: Ben Sweeting, University of Brighton, Brighton, UK. R.B.Sweeting@brighton.ac.uk

C&HK is indexed/abstracted in *Cabell's Journal* and *PsycInfo*
Journal homepage: **www.chkjournal.com**
Full text: **www.ingenta.com/journals/browse/imp**

Editorial Board

Victoria N. Alexander
Dactyl Foundation, New York

Dirk Baecker
Witten/Herdecke University
Witten, Germany

Pille Bunnell
Royal Roads University
Vancouver, BC, Canada

Sara Cannizzaro
CCSR, De Montford University UK

Bruce Clarke
Dept. of English, Texas Tech University, Lubbock, Texas USA

Paul Cobley
Faculty of Arts and Creative Industries, Middlesex University, UK

Marcel Danesi
Semiotics and Communication Studies, Toronto University, Canada

Phillip Guddemi
The Bateson Idea Group
Sacramento, California USA

Ray Ison, Applied Systems Thinking in Practice Program, The Open University, UK

Michael C. Jackson
The Business School
University of Hull, UK

Louis H. Kauffman
Dept. of Mathematics, University of Illinois–Chicago, USA

Klaus Krippendorff
Annenberg School for Communication
University of Pennsylvania, USA

George E. Lasker
School of Computer Science
University of Windsor, Canada

Alexander Laszlo
The Bertalanffy Center for the Study of Systems Science (BCSSS)
Vienna, Austria

John Mingers
Kent Business School
University of Kent, UK

Winfried Nöth
Programa de Tecnologias da Inteligência e Design Digital, Catholic University of São Paulo, Brazil

Paul Pangaro
Carnegie Mellon University
Pittsburgh, Pennsylvania, USA

William Reckmeyer
San José State University
California, USA

Alexander Riegler
Vrije Universiteit Brussel, Belgium

Steffen Roth
La Rochelle Business School, France; Witten/Herdecke University, Germany

Sergio Rubin
Earth and Life Institute
Université Catholique de Louvain, Belgium

Bernard Scott
Academician of the Int. Academy for Systems and Cybernetic Sci.

Frederick Steier
School of Leadership Studies
Fielding Graduate University, California, USA

Ben Sweeting
University of Brighton, UK

Consulting editors :

Farshad Badie; The Research Group "Natural & Formal Language," Aalborg University, Denmark

Lars Clausen; UCL University College, Denmark

Donald Favareau; National University of Singapore

Christian Fuchs; University of West-minster, UK

Florian Grote; CODE University of Applied Sciences

Claudia Jacques; Knowledge Arts Studio, New York

Markus Heidingsfelder; BNU-HKBU United International College, China

Christiane Herr; Xi'an Jiaotong-Liverpool University, Suzhou

Richard L. Lanigan; International Communicology Institute

Seiichi Imoto; Hokkaido University, Sapporo

Mark William Johnson; University of Copenhagen

Vessela Misheva; Uppsala University, Sweden

Ole Nedergaard; Copenhagen Business School

Andrew Pickering; University of Exeter

Bernhard Poerksen; Tubingen University, Germany

Devon Schiller; University of Vienna

Bent Sørensen; Aalborg University, Denmark

Torkild Thellefsen; Aalborg University, Denmark

Maurice Yolles; John Moores University, UK

Copyright: It is a condition of acceptance by the editor of a typescript for publication that the publisher automatically acquires the English language copyright of the typescript throughout the world, and that translations explicitly mention *Cybernetics & Human Knowing* as original source.

Book Reviews: Publishers are invited to submit books for review to the Editor.

Instructions to Authors: To facilitate editorial work and to enhance the uniformity of presentation, authors are requested to send a file of the paper to the Editor on e-mail. If the paper is accepted after refereeing then to prepare the contribution in accordance with the stylesheet information at www.chkjournal.org

Manuscripts will not be returned except for editorial reasons. The language of publication is English. The following information should be provided on the first page: the title, the author's name and full address, a title not exceeding 40 characters including spaces and a summary/ abstract in English not exceeding 200 words. Please use italics for emphasis, quotations, etc. Email to: sbr.lpf@cbs.dk

Drawings. Drawings, graphs, figures and tables must be reproducible originals. They should be presented on separate sheets. Authors will be charged if illustrations have to be re-drawn.

Style. CHK has selected the style of the APA (*Publication Manual of the American Psychological Association*, 5th edition) because this style is commonly used by social scientists, cognitive scientists, and educators. The APA website contains information about the correct citation of electronic sources. The APA Publication Manual is available from booksellers. The Editors reserve the right to correct, or to have corrected, non-native English prose, but the authors should not expect this service. The journal has adopted U.S.English usage as its norm (this does not apply to other native users of English). For full APA style informations see: apastyle.apa.org

Accepted WP systems:
MS Word and rtf.

Humberto Maturana (1928-2021). Photo (1997) courtesy of Pille Bunnell.

Foreword: Humberto Maturana
Reflections on Bringing Forth Worlds

Frederick Steier[1] and Pille Bunnell[2]

We are pleased to introduce this Festschrift volume of *Cybernetics and Human Knowing* written in honour of Humberto Maturana Romesin very nearly on the occasion of the first anniversary of his death (May 6, 2021) at the age of 92. Maturana remained intellectually active, wrote articles and presented seminars, all through his last years, and indeed was scheduled to present to the Cybernetics Society of the UK later that same month.

A Coherent Matrix of Ideas

In a recent note, Maturana is remembered as a polymath (Parada et al., 2021). Though he evidenced a wide range of knowledge, we prefer to recognize the value of his connected matrix of mutually coherent ideas grounded in a constitutive ontology (Espejo, this issue, pp. 63–76) This matrix of ideas, sometimes referred to as the biocultural matrix of human understanding, represents an epistemological shift that changes the way we regard "reality" as we are compelled to consider not only how we do what we do, but also how we know what we know.

As this matrix has to do with how we understand everything it could be regarded a cosmology. When asked why he did not refer to it as such, Maturana agreed that indeed it would be appropriate in the sense that it has to do with how we understand the nature of everything, but it would be rather misleading as people would hear it as having to do with the origin and evolution of the universe (pers. com., 1999). Further, it would hinder the clarity of seeing the epistemological shift he was offering.

In a special issue of *Constructivist Foundations* entitled the "Work of Humberto Maturana and its Application Across the Sciences" (Riegler & Bunnell, 2011) the foreword offers an explanation of how Maturana's matrix of ideas is substantively interlinked. All the ideas are not only mutually supportive but also mutually interdependent. Consequently, the understanding of any notion in the matrix is expanded and subtly shifted as any other notion is understood; with understanding evolving again and again as any notion is more fully understood. In practice this implies that even though a good understanding of any part of the matrix can be relevant and useful, a deeper view of that same notion can be achieved through learning further aspects of this network of ideas. For example, a variant of the matrix shown in Figure 1 was used as an outline for an MA course taught by one of us. In this course a common experience for over a decade of student cohorts was a sudden sense

---

1. Fielding Graduate University. Email: fsteier@gmail.com
2. Independent scholar. Email: life.works@mac.com

of a deeper understanding about two thirds of the way through the course. Their experience can be attributed to a shift from connecting selected notions with their existing understanding to their own construction of the coherences within the network and thus seeing it as a coherent whole.

Another way of looking at this network or matrix of notions is to think of it as an ecology of ideas (Vickers, 1968; Steier, 2005). Geoffrey Vickers gifted us with the generative phrase *an ecology of ideas*, inviting us to appreciate the mutually evolving interconnections among persons, and between their inner and outer worlds, including the relationships of their thoughts, feelings and perceptions and their surrounds. This came to be central to Bateson's *ecology of mind*, as Bateson notes in *Steps to an Ecology of Mind* (Bateson, 1972), and Vickers's own ideas of the appreciative system.

The metaphor is apt as ecosystems are not just collections of species, rather the various species interact with each other as a web of interdependencies. Furthermore, not all connections are equal, each species does not interact the same way or with the same intensity with all the other species. Rather, relationships tend to be organized in communities within any given ecosystem. Similarly, we can distinguish local densities of linkages and interdependencies among Maturana's notions. (Figure 1)

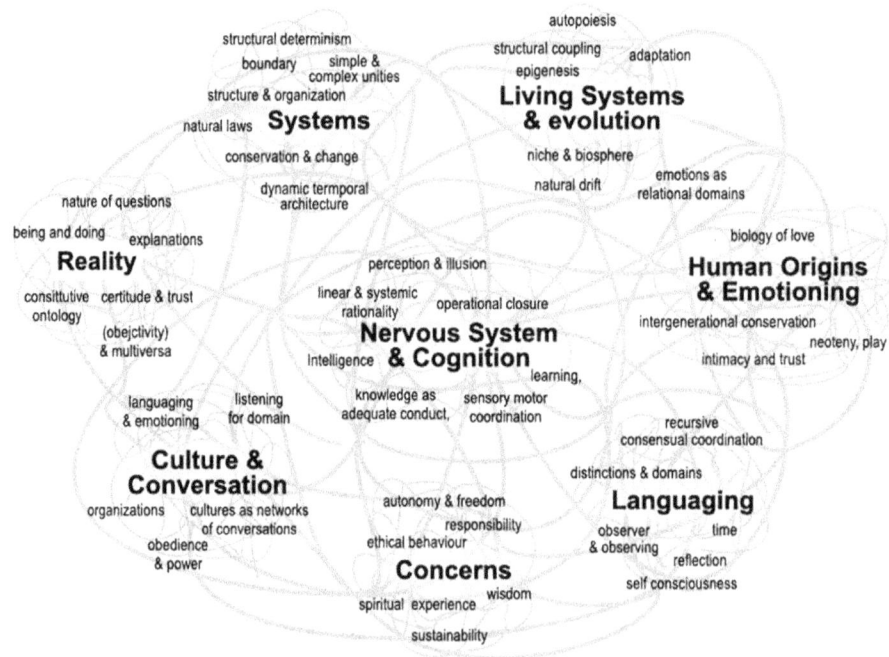

Figure 1. A Depiction of the Relationships Among Maturana's Network of Ideas With Local Densities of Interconnections.

There are a few major concerns with this representation of Maturana's network of ideas. First, the notions and their interrelations would be better conceived and articulated as a multidimensional network—though in practice it is difficult to even envision more than three dimensions. Figure 1 is necessarily constrained to two dimensions. Thus, the particular version presented here is an artificially flattened representation, with some ideas apparently pushed to the periphery and others shown as central. A three-dimensional representation would enable the viewer to regard the network from various points of view within the network where the apparent significance of relationships would shift, and appear more or less obvious as one orients in the space and, again, as one's focus of attention and concern shifts.

Second, the names given to the notions are only pointers to dynamics that must be conceptually operated to be meaningful. Third, the local collections of notions are not strict categories, at best they are pointers to denser regions in the relational space.

Finally, and most importantly, Figure 1 represents the mental framework of the person who drew it. It is one individual's particular perspective on the local densities of linkages and sense of the interdependencies and coherences among the ideas and is in no sense "real." The notion of Maturana's notion of multiversa applies also to individual conceptions of his network of ideas and no one can claim a perfect fit of their understanding with exactly how he may have conceived of these relationships. Each individual who has worked with this body of ideas has evolved, and is continuously adapting, their own networks of understanding in various ways that may include a few or many of the notions named in the figure, as well as others not included here that they attribute to Maturana. Further, even though in conversation we may experience having a mutual understanding of a given notion, our understandings may be subtly different as we have each developed our understandings in accord with the rest of our overall ecologies of ideas. Thus our understandings tend towards adequate coherences especially where the participants listen for the domain in which what the other says is valid.

Given all this it is indeed quite amazing that we are able to communicate and have a sense of understanding each other as well as we do! Languaging, in its grounding is after all a recursive consensual coordination of consensual coordinations of behavior and thus tends towards adequate coherences.

In summary Figure 1 is not "real" though there will be coherences among our understandings given that we have all developed our matrices of ideas in largely similar cultures. If this were not the case we would find no insights in reading the various papers in this issue.

The Contributions

For each of the articles included in our honoring the legacy and memory of Humberto Maturana, we offer a glimpse into their contribution. Each has also brought forth ideas that contribute to our visual depiction of the ecology of ideas as a pattern of connections for this issue as a whole that we have presented above. Since each of the articles is also multifaceted in this regard—some explicitly so, some more implicitly so, and since each also offers a rich array of possibilities, we invite you, the reader, to think of places you might locate the various essays, while recognizing that this might even encourage you to rethink what it means to locate anything in an ecology of ideas.

As the *multi*verse is a way of acknowledging the multiple turns at play in our ways of knowing and being, in contrast with a *uni*verse, so too we invite you participate in bringing meaning to the ecology of ideas for yourself, and thus, to all of us.

As a theme moving across the contributions, each of the authors has written of the ways in which the ideas of Humberto Maturana have entered into and become part of their way of knowing, acting and being in their worlds. It is a rich array of worlds that are being brought in as well, from biology to family therapy, from art to dialogue, and from education to consciousness. Each selection also offers ways of extending the ideas of Humberto Maturana both in the settings within with each works, as well as to larger contexts. In other words, the authors illustrate how they have benefitted from Maturana's work, while attempting to add to it, giving back and giving forward.

We feel it important to recognize that an appreciation of Maturana's work also involves the recognition that he was constantly inquiring—inquiring into the worlds in which he participated as well as inquiring into his own assumptions about ways of bringing forth worlds. As such there is an ongoingness to an appreciation Maturana's work that the authors are building on, working thought the ideas as we craft them, with an openness to learning about our own ideas through encounters with others. We invite the readers to read the essays in that frame as a possibility.

In "Maturana, Art and Cybernetics," Frank Galuszka crafts a mutual connection between art and cybernetics. Frank's crafting rests heavily on occasions of meetings with Maturana over the years, while at the same time bringing new meaning to those very occasions. Consistent with the artistic and painterly background that Frank brings to his writing, the essay offers rich visual images in word—and that visual aspect is a large part of what Frank offers as a treat for the reader. Frank's essay also brings forth a rethinking of ideas from Maturana and possibilities for modeling, with the section on modeling the multiverse as multistylism in art as a rich example. Frank concludes with Maturana the artist, as the performative aspect of Maturana and his idea-creating are brought to life.

Jay Efran forges a deep connection between the biology of Maturana and ways of engaging with others in family therapy and psychotherapy settings. In his essay, "Upsetting Apple Carts," Jay brings his long history of doing family therapy as well as a therapeutic educator to surface a rethinking of key concepts in therapy that derive from cybernetics and systems in therapeutic encounters. Through his own personal engagement with Maturana and his key ideas, Jay explores a rethinking of ideas of causality, information, paradox—among a host of others that are dear to the therapeutic community. What is significant is how Jay invites the reader to learn with him as he engages with this rethinking process, including how we frame the questions that we ask of helping relationships.

Juanita Brown and Amy Lenzo offer an engaging and evocative essay, "Reflections on Core Ideas of Humberto Maturana in Relation to The World Café." In their essay, Juanita and Amy reflect on the ways in which Maturana's work offered new insights for them with their development of the profoundly dialogic World Café. They note how the interweaving of Maturana's ontology of conversing, resting on his biological and cultural foundations of human existence enabled them to develop the World Café as a way of meeting committed to honoring networks of conversation and cross-pollination of people and ideas. Their casting of the World Café as a braiding of relational space with ideational space, with implications for whole systems design of our ways of being with each other is linked to core principles from Maturana's work, while at the same time, adding new dimensions to Maturana's ideas.

While many of the articles in this special issue can be seen to have rich connections, the link between Brown and Lenzo's article and that of Flavio Mesquita da Silva is particularly strong. Flavio's contribution, "Designing for Emergence: Creating Living Networks of Conversation Grounded in Love," rests on work he had done, and continues to do, in Brazil. His project, Generation of Peace, centered on use of The World Café to whole systems design and redesign of educational settings in Brazil. Flavio's essay makes clear the ways in which Maturana's ideas rooted in a biology of love and an ontology of conversing were woven into his World Café dialogic sessions and beyond, marking new ways of thinking about what it means to learn with others, about ourselves and our relationships. In so doing, Flavio also marks ways in which Maturana's ideas, particularly around the biology of love, can also be woven into the importance of conversing in a deep sense with others as a fundament of how we collectively design our futures.

Seiichi Imoto's moving essay, "The Love that was not Recommended—Maturana's Biology of Love," features many ideas that are in parallel to Flavio's. Seiichi centers his essay on a core principle of Maturana's work and life—the legitimacy of all existence. Seiichi challenges us to recognize that this fundamental notion of legitimacy of others and their circumstances is under threat. At the same time, Seiichi rests his essay on a question of why Maturana did not recommend the love that would guarantee legitimacy of existence. In what might seem a paradoxical turn, Seiichi notes it is the notion of recommending itself that needs rethinking. Seiichi offers a pedagogy of love as a way to approach this important dilemma.

Raul Espejo moves us into the worlds of organizational life. His essay, "Maturana's Path of Objectivity-in-Parentheses" invites a rethinking of relationships among organizational members, as well as a rethinking of organizations and their environments. Maturana's ideas of communication and languaging become the basis for Raul's recognition of how organizations arise through languaging relationships, and how organizational systems are crafted in networks of conversation. The reader might note parallels with Raul's ideas and the bases of the World Café. What Raul offers us as he moves us through a black box first-order cybernetics approach to his questions to a second-order approach is not just a knowledge claim that organizational systems are constructed in conversational networks, but also insights into ways of appreciating the variety of processes by which this might be accomplished. At the heart of Raul's approach we find Maturana's objectivity-in-parentheses as inviting a constituted objectivity of the reality of organizational systems.

Objectivity-in-parentheses also serves as a key point for Hugh Gash's essay, "Humberto Maturana: Using his Biological System in the Social Domain." Hugh traces his own learning about learning and teaching, bringing Maturana's ideas rooted in biology to his constructivist and cybernetic approaches to education and ways of knowing rooted in Dewey, Piaget and von Glasersfeld. Hugh focuses on how he has brought learning about topics in school settings where different perspectives are deeply rooted, such as gender stereotypes, to questions of what we might want of tolerance, and whether that is even enough. In moving from bringing these questions to educational settings to ones of climate change and sustainability, Hugh brings in Maturana's objectivity-in-parentheses as a resource for social learning, and community dialogue in creative ways.

Kathleen Forsythe poetically weaves in her learning from the ideas of Maturana and her occasions in meeting with him. She does so within the context of her own work with youth who are non-speakers diagnosed with autism. Kathleen's essay, "How My Understanding of Languaging in Non-speaking People With Autism Has Been Informed by Conversations With Humberto Maturana," invites us to think about language, languaging and communication in different ways, and how those different ways allow us to recast our assumptions of what it means to participate in world-making with others. Kathleen weaves conversations over the course of her career, with both Humberto Maturana and Pille Bunnell into her own learning, and the lives of others. In so doing, Kathleen also stresses the importance of emotioning as a fundament of how Maturana's biology of love extends to the social realm, as well as what it means to be present for and with others.

In his essay, "Consciousness: An Unresolved Question, and What Maturana Has to Say About It," Pier Luigi Luisi invites us to consider the relationship between cognition and consciousness. He does so by bringing Maturana's work into the question, while also pointing out the difficulty of comparing Maturana to other authors. Indeed, the basis for making the comparison difficult for Pier Luigi is the fundamental starting point for Maturana of human existence as occurring in "the relational and operational space of living" (Maturana, 2005, p. 73). Pier Luigi's essay encourages us to, at another level, how we bring Maturana's ideas into worlds whose boundaries themselves may need re-examination, as we think about our own existence and that of our planet.

Jude Lombardi and Larry Richards offer an energizing dialogue with each other as they ponder meanings of Maturana's proposition of a zero-time cybernetics. Maturana had proposed a zero-time cybernetics on his receipt of the Norbert Wiener medal for lifetime achievement from the American Society for Cybernetics in 2008. The letter he sent in recognition of receipt of the award serves as a background for the evolving dialogue between Jude and Larry about what *zero-time* might actually mean. Consideration of how cybernetics, rooted in cyclicity, might inform such an idea serves as a backdrop for their dialogue, which can be read as invoking, in their process, the very ideas they are in dialogue about. Their essay concludes with provocative questions that can be seen to emerge from their dialogue, including paradoxes of how we live our lives in time day to day, while at the same time recognizing how we invent time.

Lloyd Fell enriches our special issue with a short essay, "Animals and Humans Alike." Lloyd's essay encourages us to recognize how Maturana's ideas on biological systems need to be recognized as about all living beings. Lloyd's starting point is with farm animals, but his brief essay can be seen to offer the importance of Maturana's ideas to inter-species communication and understanding, and how key principles of autopoiesis and structural coupling need extend to all living creatures and our relationship with them and with each other.

The concluding article of this section of our special issue is by Bernard Poerksen, as he offers a chapter from the book that he and Humberto Maturana published, *From Being to Doing: The Origins of the Biology of Cognition*. This selection in our special issue entitled "Power Arises through Obedience" contains a context setting introduction by Bernard, making clear not only the ideas that are brought forth in the essay, but also the importance of this conversation today. The chapter itself is a rich dialogue about Humberto's life during the dictatorship in Chile, and how such experiences shaped his ideas. The reader is invited to recognize the systemic wisdom at the heart of Maturana's work as living and learning in difficult circumstances are brought together in a deeply reflective dialogue.

Our special issue also contains regular features of *Cybernetics and Human Knowing*. We are pleased to present our regular journal column written by Bernard Scott. His essay, "Second Order Cybernetics and the End and Beginning of Philosophy," builds on an earlier guest column of his from 2019, "In Defense of Pure Cybernetics." His current essay seeks to bring philosophy back to its roots—and etymology—as a loving of wisdom, through a weaving of second-order cybernetics into our processes of knowing and being. As many leading philosophers and cyberneticians are woven into the mix by Scott, the braiding of second-order cybernetics and philosophy allows for an emergence of an art and science of fostering goodwill.

We also have the American Society for Cybernetics (ASC) for this issue. Ray Ison has taken the occasion to offer a special ASC column that fits with the theme of this issue, as he offers reflections on personal encounters with Humberto Maturana over the years. Ray crafts his column, "Mutual Arisings: Conversations with Humberto," as an extended dialogue revealing the ongoingness of his learning about Maturana's work, and his own work through Maturana's ideas, over thirty five years. What is fitting is how it is clear that even writing this essay has extended Ison's learning with Maturana.

Finally, we have a book review by Mark William Johnson, entitled "Leydesdorff's Compass." Johnson offers a review of communication and systems theorist, and cybernetician Loet Leydesdorff's recent *The Evolutionary Dynamics of Discursive Knowledge*. Johnson's comprehensive review provides rich insights into the huge scope of Leydesdorff's work, taking care to offer us the metaphor of a compass to guide our journey.

Artists Statement

The artist for this issue is one of us guest editors, namely Pille, so what follows in this section is in her voice:

It is a special privilege to be able to offer my photos as the artwork for this special issue and to do so as a further homage to Humberto Maturana. He had a deep appreciation of the beauty in the natural world and expressed the opinion that such appreciation is not only a human faculty but is shared by other animals. To me that makes sense, as the experience of beauty seems to be the consequence of seeing coherence, or fit, with and within one's medium. This in turn seems to be a natural consequence of living always happening in the dynamic of structural coupling, a process that both creates and thrives in coherences. It is only the reflective pause that we humans excel given that we live in a languaging niche. Sometimes it is immensely refreshing to abandon the world that arises in language and experience in a manner that feels more direct. Most of my images are derived from such moments. They are a response to my own encounters with the natural world, to those moments when perceived harmonies override any sense of naming or classifying; of considering anything other than an evoked sense of beauty.

Yet even as I claim this as an experience, I remain aware that even that which is experienced without any sense of language is also in recursive structural coupling with the languages (idiomas[3]) I have lived in.

Given my awareness that the human brain has evolved and developed in the dynamics of living in language, the way of being present, without consideration of naming or distinguishing, that gave rise to these images is after all not incongruous with the requirement to provide a name for each published image. In the event, I have named some of the images by what we socially agree they represent, whereas I have named others by alluding to some particular notion from Maturana's network of ideas. I found that I could create an imagined congruence that enabled me to map a dynamic notion to a static photo. This allocation of Maturana's notions to images feels rather presumptuous, yet it is an interesting exercise to consider what one distinguishes in a visual derived from some glimpse such that one can attribute a notion in language to that image. Better yet if in that mapping one can explore a further nuance of the notion itself, as was the case for me. Thus, even as we invited you to consider your own networks of understanding as your own evolved ecology of ideas, so we would also like to invite you to consider and reflect on whatever the images evoke for you.

Finally, I wish to state that I appreciate the opportunity to include several images that evoke some of the many aspects of my beloved friend and mentor, Humberto.

---

3. In Spanish there is a nice distinction between language as what we humans do (lenguaje) and the various languages we have evolved (idiomas). Here I mean the latter, as I think differently in the *idiomas* I know as a native speaker.

Acknowledgement

We wish to acknowledge the help of Jeanette Bopry in preparing this issue. Jeanette is further acknowledged for her various editorial capacities going back many years, including her present position taking on the responsibilities of an editor in chief. Her contributions to this issue, and to many previous issues, extend well beyond the normal roles of editor or associate editor, to helping bring about a context of learning at so many levels. Thank you, Jeanette!

References

Bateson, G. (1972). *Steps to an ecology of mind*. New York: Ballantine.
Maturana, H. (2005). The origin and conservation of self-consciousness: Reflections on four questions by Heinz von Foerster. *Kybernetes, 34*, 54–88.
Parada F. J., Rossi, A., & Rojas-Libano, D. (2021). Chilean polymath Humberto Maturana remembered. *Nature, 594*(7862),177. Doi: 10.1038/d41586-021-01524-8.
Riegler, A., & Bunnell, P. (Eds.). (2011). The work of Humberto Maturana and its application across the sciences. [Special issue.] *Constructivist Foundations, 6*(3).
Steier, F. (Ed.). (2005). Gregory Bateson: Essays of an ecology of ideas. Cybernetics and Human Knowing, 12(1-2)
Vickers, G. (1968). *Value systems and social process*. London: Tavistock.

Bunnell, P. (2020). *Poppy Sepal*. Photograph.

Bunnell, P. (2007). *Zero Time*. Photograph.

Maturana, Art and Cybernetics

Frank Galuszka[1]

This paper explores links to art, particularly the art of painting, to the work of Humberto Maturana. It looks at the long-standing role of art in cybernetics and the role of cybernetics in art, and looks toward potential for future creativity and discovery in both fields. As well as his ideas, Maturana himself, as a man, a performer, and as an artist, is affectionately remembered in the context of this paper.
Keywords: Humberto Maturana, Abstract Expressionism, metastyle, multistylism, multiverse, dissociative identity disorder, multiple personality disorder

---

I. St. Gallen[2]

We do not come cold to ideas. We are already warmed up. In my experience, I encountered Maturana in a cold place however, Saint Gallen in snow, high on the hill, in the chilly lecture rooms of the university, with modernist wood furniture, concrete walls stained with damp, Maturana himself in a warm hat and a scarf, lecturing thus, drawing on a chalkboard, first an eye, high up (his arm reaches with the chalk) on the left, "There must always be an observer," then, below its gaze and to the right a pair of deformed circles, interrupted, and each with an arrow suggesting movement, "Living systems," followed by a deformed line below them, "a medium," then back-and-forth arrows between the living systems and more arrows back and forth between the living systems and the medium, followed by a little erasing and adjusting of the lines so that it is evident that the livings systems and the medium are all beginning to conform with one another.

The hat is a rustic grey-brown felt hat, such as a shepherd might wear. On another mountain. The sound of the chalk on the blackboard.

Done, he turns around, nimbly and elegantly to face a split audience, half on his side, and half against. He raises his charm and charisma with perfect timing, but I am already captivated. By the diagram. By the diagram itself.

Certain things are visually true. True at first sight. This diagram on this blackboard with this audience in this cold with this man, with his hand turning at the wrist as it draws on the board, with the hat he is wearing, and the scarf haplessly trailing, is suddenly a unity that is true. Before he speaks I understand this

Yes, there must always be an observer. That there must always be an observer is something that I know.

---

1. Professor of Art, University of California, Santa Cruz. Email: frgalusz@ucsc.edu
2. I have referred to several conferences in this article as occasions where Maturana's talks and interactions were prominent. These include ASC conferences at St. Gallen, Sundvollen, Seabeck, Philadelphia and Virginia Beach; a lecture at the University of California, Santa Cruz; and a workshop at Horsham Clinic, Horsham, PA.

Given a full day at this conference, my listening ears are already brimming with things I don't understand. Other diagrams and talks that seem upside down but fleetingly true, as in grasping one idea I lose sight of the one just before. A diagram by Fred Steier of things inside things, open here, but closed there. Before and after Maturana, the down-the-rabbit-hole teapot mystery of furtive, caped Gordon Pask, the suave mechanics of Heinz von Foerster, other luminaries and glancing lights, nestled in the diffractive abundance of snow. What could be better?

As Maturana spoke, I'd decided already what side I was on, but was glad of those that were against him, who (fighting the cultic allure as well as the content) posed questions I might have posed if I'd had the wherewithal to formulate and courage to ask, with each of the provocations building, often elegantly asked, yet equally elegantly answered. No defensiveness, not really. Maybe a little impatient, frustrated at being misunderstood. I looked at the cold high concrete walls, the blunt modernist architecture, the wide blackboard, the diagram, the man, his hair, his glasses, his stature, his movements, the hat, the scarf, the voice, the worn nub of chalk still in his raised hand as he spoke.

After his then controversial lecture, with the tension still high, the room settling into camps of elation and grumble, Kathleen Forsythe, putting on her coat, asked me, "Is this science or religion?"

II. Maturana's Blackboard

There was the drawing on the blackboard. I wasn't interested in arguing with it. The drawing on Maturana's blackboard I took as a universal diagram for all examples of living systems with one another in a medium. Any medium. Any of many. It jumped out of the thinking world into the world of living.

It filled every space. Every corner of every space, thing, and thought, it operated its connecting living machinery, to systems in nature, social situations and to art. It applied to all aspects of art and to the art of painting: painting as a process, painting as psychological, painting as a form of communication: communication between the painter and the viewer, communication between the painter and the painting, communication between the elements within the painting, communication of individual brushstrokes to each other and to the medium of the painting, communication between the painting and the body of work of which it is a member, communications between communications within bodies of work to one another, correspondences, forming and deforming mutual influences, communications between the art of an artist and the art of a culture. The elements of a painting as a living system within the period of its creation, with each added element being responsible to, and influential upon, every mark, gesture, decision, relationship. Every painting was energized with living links, and every painting and every part of every painting was linked to every other in a changing, surging, informational, emotional, strengthening, weakening, tightening, stretching, but nowhere in a disconnecting way.

III. A Studio

A studio. Paint-spattered floor. A space heater. There are two models posing. They are in costumes as though they are characters in a play. These two models are best friends. Facing each other they hold hands, left in right and right in left. Staring into each other's eyes, they sing to each other. An observer, holding a brush to a canvas, observes two living systems, in a medium, singing. Singing to each other.

IV. Style

Like cybernetics, painting accepts the legitimacy of paradox. A painting is a whole system. At the same time, any painting is nested in histories of production, the more or less distinct contexts that bound it. Contradictions that would consign certain trains of thought to non sequitur, exist in painting (and cybernetics) without contradiction though with *dynamic logic*, high energy and considerable friction.

The most visible, disconcerting and dramatic impact of Maturana's view is its consequences to style. Style is the conscious or partly conscious outcome of artistic sensibility, the system of allowables and disallowables that any artist applies to each work of art.

Maturana's *multiversa* is an aggregate of constructed universes. While I occupy my constructed universe, I am invited by Maturana to consider that others similarly occupy universes of their own construction, but that these universes are not definitive or accurate in that none of them can correspond to the imaginary universe-as-it-is. Style, in art, is a model of a constructed universe, whether intentional or unconscious. By style we see that a Monet is a Monet, and that a Giacometti is a Giacometti.

Style is so associated with the personality of the artist that it is identified with it. The code of allowables and disallowables that underlie the choices that result in any style have come to be identified with the personality of the artist who produced the work.

Peter Schjeldahl (1990) regards painting as an unbeatable metaphor for individual consciousness, and also as the best vehicle for individual consciousness. I agree with his view.

V. Metastyle and Multistylism

But, in my work I had stepped out of my self-constructed, albeit already rebellious style, to generate a concurrent offshoot, that was sealed, comprehensive, and independent. This creation created a crisis, not only of my style, but of what serious art could or could not be. The rogue breakaway style violated a longstanding critical rule of unity, so longstanding that it was unquestioned. Some years afterwards, at a Maturana workshop in Horsham, Pennsylvania there is a patient and his family on closed circuit TV. The patient has multiple personality disorder (dissociative identity disorder). He exhibits several *alters* in the course of the interview with the therapist.

In terms of Maturana's diagram, I wondered, where is the living system? What is the medium? If at one level living systems (and mediums) are cohabiting with other living systems in a medium, then, in terms of adequacy for survival, the unity of personality, equal to the unity of style, is not a necessity. What could function in life could function in art.

A metastyle, or meta-medium, could lie outside the individual style (including its history and future) of the artworks and even outside whole bodies of work, and that the art, in its deepest parts, is in their interactions with it, and that every new artwork changes the configuration of the medium. Each painting speaks, unseen, to all the others, and across stylistic barriers, and this is the greater part of the art.

With the elimination of a master style or overarching concept as conscious organizing principle, the condition of the multiverse could be modeled as multistylism in art.

Within such an unconscious metastyle, a multistylism constituted of closed-system styles, could sustain itself as analogous to dissociative identity disorder sustains itself through an unseen organizing principle, and as analogous to the multiversa.

VI. Modeling the Multiversa as Multistylism in Art

My understanding of how style advances in art, with the art of painting in mind as my example, requires a complaint, a continuation, and an outside influence.

There is a dissatisfaction or unrest about the style of art that dominates the cultural environment. Something is wrong with it. Deficient. This complaint prompts a survey, including a survey of past art, for something that seems better suited to the present. At the same time there are also parts of the art of the present that we wish to continue, to carry forward. Third, there is an outside influence: This can come from another culture, from something that hasn't been thought of, hasn't been considered, has been discarded, is commonly disapproved, or is thought to be unthinkable in art.

My education in painting had been in the abstract expressionist period. By the time I encountered Maturana, my complaint had drawn me into figurative painting, because of a list of arguments I won't go into here. What I carried forward in my work was the abstract expressionist ethic about what it was to paint. To paint with integrity. (Painting concerns itself with integrity.) While my work had undergone waves of outside influences, Maturana, and cybernetics as a whole, presented a wave of another order.

From the abstract expressionist perspective, a painting is not only an object, but foremost, it is an act. Whether selections and encodements can be examined from the final product is not the point.

Paintings are partly conscious and partly unconscious creations. Paintings, like other things, come about through a coordination of coordination of actions. In the case of the painting, its existence as a closed system is straightforward, obvious. Whether the actions that created it are substantiated as decisions that have a quality of awareness of responsibility to each other and to the whole, is another matter.

Maturana's thoughts suggested a way to expand the integrity of painting, and its reach of style and content through acknowledging the functions of living systems within mediums and the multiverse.

The Abstract Expressionist ethic held that a painting is not finished; it is abandoned. It is to be revised toward its emergent identity, but never designed into it. The emergent identity is elusive. A certain silence, maybe, a presence. Self-consciousness is to be suspended. The painter exists inside the act. There are periods of action followed by periods of reflection; reflection leads to resolution regarding a next corrective action, hence: critique, critique, critique, revise, revise, revise. Thus, the ethic was to paint always into a crisis, not away from it, not towards a finished work.

In this approach, the sequence of decisions can be seen as a forceful, narrow path forward, a fire hose of unconscious effort. It excludes all else, in seeing-feeling the visualized identity as an unconscious emergent, and in teasing it into reality. Revisions aim at sweeping noise from the path to the signal, at the cost, eventually, of the sacrifice of much that has been gained. "My work is the sum of its destructions," Picasso said.

Maturana's coordinations of coordinations of actions model, attached to this ethic, expands it (this is a metaphor) from a harsh high-pressure channel to a blooming volume edged by further-expandable frontiers of questions and newness. This preserves the ethic of working toward the point of crisis, and through it, and again, not retreating into a finished work. My immersion in metastyle and multistylism has had several phases. Immersion in a metastyle produces contact with an unknown of a different order in art, and a heightened level of crisis.

VII. Cybernetics and Art

At an ASC conference in Virginia Beach in the late 1980s, Larry Richards asserted the significance of the arts to cybernetics and of cybernetics to the arts by beginning the conference with musical and theatrical performances, and by inviting Glen Davidson and Anne Hayes to construct a giant Paskian double torus during the course of the conference. Art became a topic of discussion. Postmodernism was in full sway at the time of this conference, and issues like authorship, originality, style and even significance were being held in question. How best could art reflect our times was debated during a late session. This conference led to another about a year later in Montreal. Davidson and Hayes built a giant paper thistle, and presenters and performers merged together.

Cyberneticians in general enjoy crossing over into nonverbal unknowns. So many are artists, and these are so diverse. Brun, Krippendorff, Pangaro, Parenti, Beer, Pask, Pedretti, Forsythe, to name only a very few.

About this unknown, Pask was especially canny, presenting a talk, *Against Repeatability and Reliability* (Pask, 1989) to legitimize it. Further, he insisted in an ASC 1995 conference in Chicago (November 17-21, 1995) that we should not forget that, with whatever it is we are considering, "there is always a black box." And von Foerster, a designer himself, reminded us that "we do not see that we do not see," and how rarely it is that things are truly undecidable and that it is only among truly undecidables that we are free to decide. Marcelo Pakman studies the near impossible, from what feelings are, to the succession of uniquenesses, or singularities that make up our moment-to-moment experience. It is in art that *singularities*, undecidables and black boxes are lent frame and form, and in art that we formally and knowingly enter into them. Knowing we have stepped out of, and into.

VIII. Maturana as an Artist

Maturana has called cybernetics the art and science of understanding. During the memorable 1989 conference in Virginia Beach, he said, toward the end, almost offhandedly, "While we live we are immortal." This assertion, paradoxical and penetrating, is, for me, at the center of his view, and at the center of his own understanding.

His brother was a painter and urged him to paint. Maturana, according to my memory, made one painting and decided to paint no more. He reckoned that, if he continued, he'd want to paint every day.

He was attracted to fiction. Fiction is the what, that is not here, becoming something that is. He said he liked Nietzsche because Nietzsche told stories. This reminds me of Gregory Bateson, who was always inclined toward stories.

Sometime in the early nineties, when, as a correspondent for the *Philip K. Dick Newsletter*, I was tasked with interviewing Heinz and Humberto, I asked each about third-order cybernetics. Heinz responded that no, it will always be second-order cybernetics—that any meta-cybernetics has second-order status, while Humberto responded, "second order, third, as many as you like." To me, this difference of response, showed a difference between these two creative men, and it leaves me wondering what kind of fiction Maturana had in mind when he entertained the possibility of third-order cybernetics. Am I wrong to believe that Heinz's impulse was toward containing, and that Humberto's was toward opening up? Is this the difference between the engineer and the biologist? (And I must say here admire Judith Lombardi as she persists in inviting an imagined third order into serious consideration.) As the second order had been earlier, the third order is felt then and now, like a paradigm-shaking prophecy that can be made real if we wish it. And the desire, whether it be called one thing or another, is to create an equivalent productive rupture as second-order cybernetics had made, fulfilling the desire for yet another new world of thought and action to spring from cybernetics.

The signal or presence sought in painting, which can be tested for authenticity only through sensibility is comparable to what attracts some of us to an imaginary third-order cybernetics. It floats like a grail in the imagination, as a waiting gift, an access to greater understanding, tying heightened and liberated imagination to a hidden, yet intuited coherence.

In a conference in Sundvollen, in Norway, Maturana said he is trying out something he had been thinking about. It was a description of love, later to be called the *biology of love*. In this presentation, he enacted a performance, a miniature drama. He specified a legitimate other as a spider, and pronounces the word *spider* with relish, enjoying especially the split of the syllables at the middle of the word, framing and focusing the sound so that a spark of attention flashes there. He performs how it is he will act in regard to the spider, approaching its imaginary place in a walking gait, seeing it in his path, pausing, looking it in the face, nimbly dancing around the little fellow and gesturing appreciatively toward it, before resuming his path. He shows, by giving the imagined spider such detailed regard, that the spider is fully a legitimate other of stature and dignity comparable to his own, that both he and it are equal and bonded in the biology of love.

I prefer the talks and the performances to the papers and books. In this early iteration the fullness of presence of the self and of the other were vividly comparable, the fullness of one and the fullness of the other—each *other* as a legitimate other in relation to one's self, later became disappointingly compressed as *in relationship*, which, for me, whisks away the very poignancy of relationship it seeks to describe. It falls from art into science as it folds individual expression into a generalized contract, eliminating the discovery of relationship in the moment extracted from the flow of living. Furthermore, the early forcefulness and freshness of this realization deteriorates into a concept-object over time, and to be used in a functional toolkit of concepts with which to argue further things.

In a lecture addressing the distinction between perception and illusion, he proclaimed that, while ghosts do not exist, everyone knows well the characteristics that would distinguish a ghost. He goes on to become specific, saying, there are eleven ghosts in this room, and adds details. Why does he find it necessary or desirable to make a vivid case for what he claims does not exist? He enjoys the fiction, and enjoys the kind of reality that fiction is. He enjoys that, in fiction, that which does not exist does exist. And everyone knows it. Shares it. Paradox itself is something like a ghost, a folkloric entity that is a nonentity that is an entity of a sort that gladly clings to cybernetics.

Among his informal discussions, his Q and A's, there was one memorable and long lasting at Seabeck where he discussed with candor why he remained in Chile during the Pinochet regime, and how far he was personally willing to go, as to medical interventions, drawing the line at organ transplants. Later at UCSC describing cheating as requiring expanded creativity. I find his candor illuminating and tender.

When Humberto, myself and some others were recruited to start a stalled car by pushing it, and succeeded in helping it get started again, he exclaimed, "What powerful horses we are!" and, with that, we felt sudden pride and delight in our momentary group membership. He often used the term *cozy*, and there is coziness in intimate membership and in the imaginary of shared fiction. I wonder if shared fiction is not at the heart of what religion, from a cybernetic perspective, might be. Living in metaphor. With or without transcendence as an explanatory principle.

When, in any lecture, Humberto presents his diagram of living systems in a medium, it is not only a diagram, it is a drawing, a work of art, executed for us with full and generous lines, and with the cozily familiar observing eye at the upper left, not only indicating, but expressing a being with a consciousness.

I invited Maturana to speak at the University of California in Spring of 1997. His talk was open to the whole university and was in a large hall. Since he had been invited by the Art Department, he decided to speak on something that was connected to art: aesthetics. Little did he realize how unpopular, by that time, aesthetics, as a topic and as a distinction, had become. He discussed animals at twilight. Animals in groups. He talked about their behavior in response to the change of environment as the light lowered at the end of the day.

One or two of the leading lights of the UCSC faculties stood up and walked out. Over the next five minutes others followed, diminishing the audience.

Maturana often asked, in examining an action in a living system, "What is being conserved?" What was being conserved by these academics when they made a show of leaving this lecture? I have my thoughts but I'll save them for another paper.

I've loved Maturana's work in performance, as the art of cybernetics. I feel that without the infusion of life and, even, without uncertainty and ambiguity, the art of cybernetics is not understood. One of the prime characteristics of cybernetics is transient understanding. What is unknown becomes, for only a moment known, and then slips back into being unknown again. These understandings are not portable. They appear as epiphanies. So it is that, for any participant, cybernetics conferences unfold processions of revelations that cannot be taken home in a pocket, a notebook or even in a recording. Cybernetics's necessary double existence, as art and as science tips the scales this way and that. The science of it conserves credibility and dignity, assures instrumentality, while the other, the art of it, is of another order. It could be said that grace has a part in it. Humberto had a lot of grace. The grace to take not infrequent arduous flights, changing planes in Miami, to talk, the grace to endure fawning fans and hecklers, and walkouts, to be deluged with friends, colleagues, students, groupies and hangers on, the grace to eat strange food, at strange hours, to be consistently nice to hosts, funders and waiters, the grace to speak as if fresh and fully present while suffering with exhaustion and jet lag, to large or small, adoring or indifferent audiences.

I like his work in the spoken form. In informal forums. In Q and A's. Spontaneity, improvisation, before arriving in publication chafed by translation, frozen in type, descended into doctrine. There is a taste for unfinished artworks. I have it. The finished state is in sight, while the means of its creation are vigorous and bold from search and from the urgency and uncertainty of capturing a vision. Again, "What is being conserved?" The poet is an adversary of immobile precision as poetic perfection is otherwise, as it needs conserve the spirit of its creation.

IX. Conclusion

It's late in the day. A creaking floor. A ladder. The two best friends, are standing by the heater in the studio and holding hands are singing to one another. A tall painting nearly reaches to the ceiling. I apologize for its long title, *La Vista Totale, a partial view: Between Amherst and Delphi*, but the painting has all of this in it. This painting wouldn't exist without Maturana.

And so, in answer to Kathleen's question about the lecture, as she puts on her long lavender quilted coat to prepare for her return to the cold on that snowy day, "Is this science or religion?" It is both, of course, I know she knows, and art, of course, also.

Reference

Pask, G. (1989). *Against repeatability and reliability.* Presented at Learning and Designing: A Workshop, The Center for Cybernetic Studies in Complex Systems at Old Dominion University, Virginia Beach, VA, April 21-23, 1989.
Schjeldahl, P. (1990, November 13). The new low. *The Village Voice*, p. 97.

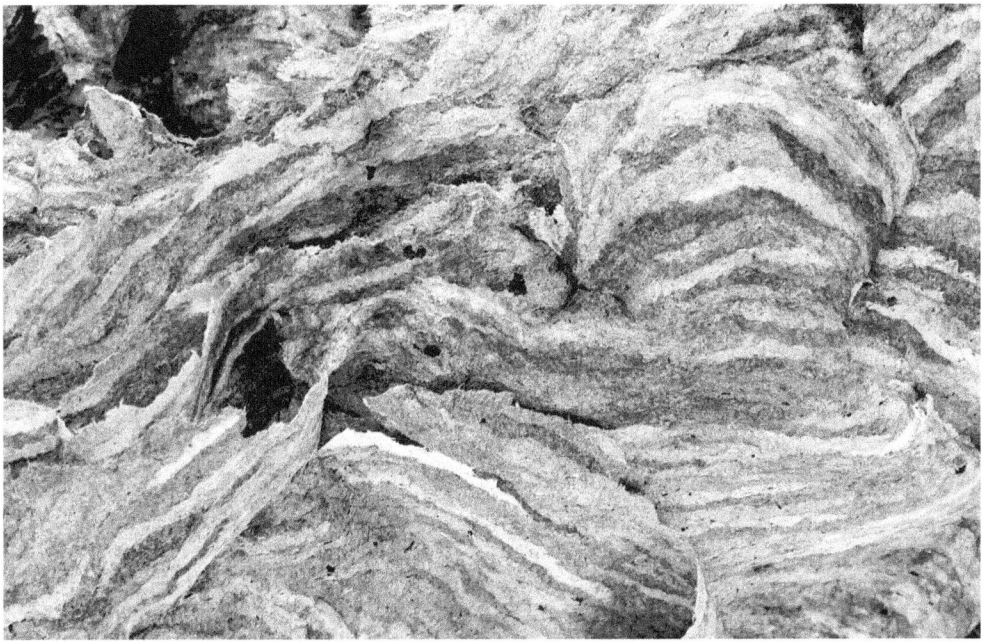

Bunnell, P. (2020). *A Bit of Niche* (wasp nest). Photograph.

Bunnell, P. (2016). *Perception & Illusion*. Photograph.

Upsetting Apple Carts

Jay S. Efran [1]

Humberto Maturana's theory of structural determination has had major impact on the field of family therapy and, later, on the work of therapists who see individuals and couples. The field first learned about his ideas in the early 1980s through a series of articles by Paul Dell, who reported that Maturana's perspective raised important questions about the validity of some of family therapy's most cherished notions: homeostasis, resistance, paradox, open and closed systems, information, hierarchy, and causality (both linear and circular). This came at just the right time to curb some of the field's growing misinterpretations of systems thinking. In the current piece, I detail ways in which Maturana's work clarified my own thinking about psychotherapy, including modifications in the way I view the notion of joining, the issue of therapeutic hierarchy, the definition of *cure*, the relationship of insight to change, the dynamics of emotion, the definition and role of language, the nature of client complaints, the significance of framing the right questions, and the importance of understanding life as a natural drift.

Keywords: Humberto Maturana, structural determination, family therapy, cybernetics, emotion, natural drift.

In my opinion, some of the most useful books about psychotherapy have been written by non-therapists. These are books about how we live, acquire knowledge, and navigate the world. The writings of Humberto Maturana fall into this category, along with Gregory Bateson's *Mind and Nature: A Necessary Unity* (1979) and G. Spencer Brown's *Laws of Form* (1972). When I was teaching, I regularly put *Mind and Nature* on my reading list, and I frequently quoted from *Laws of Form*. However, I knew better than to directly assign Maturana, for fear of precipitating a classroom mutiny and perhaps an unpleasant trip to the dean's office.

Part of the value of these authors is that they did not have to worry about the suppositions of the therapy community or the tenets of the medical model. For instance, in a 1985 interview, Richard Simon (the editor of *The Family Therapy Networker*) specifically asked Maturana why therapists should care about his work, given that he is a biologist who studies such things as the color vision of pigeons and the retinas of frogs. Maturana's reply: "I don't know. That is their responsibility, not mine" (p. 36).

Similarly, at a 1988 conference at the Horsham Mental Health Clinic in Pennsylvania, a therapist-in-training recoiled at Maturana's suggestion that all patterns of living were biologically legitimate. Why then, she asked indignantly, should she bother to become a therapist and devote her life to changing people? Maturana's reply was curt but accurate: "Because you want to."

I suspect the young woman was expecting Maturana to mouth some platitudes about the importance of being a mental health professional. Instead, he placed the

---

1. Temple University. Email: efran@temple.edu

responsibility for her career choice squarely on her own shoulders. Like most of what Maturana offers, his response was crystal clear but potentially irritating, upending lots of proverbial apple carts.

General Semantics, Communications Theory, and Clinical Training

My own interest in language and patterns of communication precedes Maturana, harking back to an elective course that I took as a graduate student. The topic was General Semantics, and the course was based on Wendell Johnson's (1946) classic text, *People in Quandaries* (1946). I found it to be an eye-opener and wondered why this material was not being taught in any of my required classes. Years later, I was equally baffled by the failure of most clinical training programs to assign works such as *Pragmatics of Human Communication* (Watzlawick, et al., 1967) or *Change: Principles of Problem Formation and Problem Resolution* (Watzlawick, et al., 1974).

I first heard Maturana's name mentioned in a 1975 conversation between Werner Erhard (the originator of est training) and a member of his staff. I overheard them say something about a Chilean biologist whose thinking would revolutionize our understanding of science and life itself. However, because I was essentially eavesdropping on their private conversation, I was in no position to ask for clarification. Thus, it wasn't until seven years later, when I read Paul Dell's articles on structural determinism (1982a, 1982b, 1985), that I began to understand what the fuss was about. At around that same time, I saw an announcement indicating that Dell was going to present at an upcoming Family Therapy Symposium in Washington, DC. I immediately made plans to attend. But, at the appointed time, the conference staff announced that Dell had cancelled—he was in bed with the flu. Because so many of us had travelled to Washington specifically to hear him speak, Richard Simon—the conference organizer—called him at his home in Virginia and persuaded him to get out of bed long enough to placate his many fans. Dell valiantly made the trip to Washington and—armed with a box of tissues and a fistful of aspirin—regaled a packed room with his take on Maturana's views.

In that era—the late 70s and early 80s—family therapists were already being seen as the mavericks of the mental health field, having challenged the all-too-American notion that psychological problems are housed in individuals rather than relationships or groups. This was a time when psychoanalysts, concerned with preserving the sanctity of the consulting room, rarely agreed to speak with a client's family members. Non-analysts were a bit less tight-lipped but no less wedded to individual treatment paradigms.

In addition to the "original sin" of working with more than one person at a time, some family workers began to make use of the perspectives being pioneered at that time by cyberneticists and communication theorists. Yet, in their enthusiasm for systems thinking, many went overboard. It was certainly reasonable for them to relabel the patient as the *identified patient*—a reminder that he or she was not necessarily the root cause (or sole container) of the problem. However, they also began reifying the family system, as if it had an existence independent of the family's collective activities. In some quarters, it became de rigueur to portray the family system as a clever enemy that resists change and must be outfoxed by the therapist through the use of various strategic ploys. From this perspective, symptoms were construed as often having disguised (systemic) functions. For instance, a boy's refusal to go to school might be interpreted as his crude (and perhaps dimly understood) attempt to protect his mother from unacknowledged spousal abuse. During this era, systemic therapists became enamored with concocting various paradoxical injunctions and devising elaborate *reframes* to disrupt the status quo and overcome so-called system resistance.

One of the wilder versions of this sort of intervention was Mara Palazzoli Selvini's *invariant prescription* (e.g., Selvini Palazzoli, Cirillo, Selvini, & Sorrentino, 1989, p. 221). She might, for instance, advise the parents of a psychotic adolescent to leave home for the evening (or longer) without warning or explanation. When they returned, they were not to explain where they had been or why they had left. This maneuver was calculated to destabilize the presumably maladaptive family system and trigger a healthy family reorganization. There was precious little hard evidence that these strategies were effective.

Of course, not all **family** therapists subscribed to these practices. Many of us considered such methods excessive and manipulative. It seemed particularly unwise to lie to clients and knowingly peddle false interpretations of their problems. Even though it was well-intended, promoting fabricated explanations hardly seemed like the best way for a profession to garner public trust.

Thus, many of us felt that Dell's descriptions of Maturana's perspective came along at just the right time to rein in the enthusiasm of family therapists for some of these dubious practices. Dell argued that systems were, in fact, not ultra-devious entities with which therapists needed to do battle. Applying Maturana's ideas, Dell questioned the rationale for seven of the family field's most cherished concepts, namely "homeostasis, resistance, paradox, open and closed systems, information, hierarchy ... [and] the whole idea of linear and circular causality" (1982a, p. 40). He contended that "the idea of a family maintaining the status quo by resisting change is, to use Bateson's phrase, an epistemological error" (p. 40). Moreover, although agreeing with the field's skepticism about simplistic cause-effect thinking, he noted that the kind of circular causality they continued to champion was "every bit as erroneous" (p. 41). As Dell explained, the problem does not lie with whether the causality is considered linear or circular, but "with the idea of causality itself" (p. 41). In other words, from Maturana's perspective, Dell said "interactions do not involve Newtonian efficient causation, but rather a relativistic structure determinism" (p. 41).

Shortly after reading Dell, I happened to be chatting with several therapists at a nearby mental health center. In keeping with the jargon of the day, they were talking about how they had "joined" the families they were seeing. It immediately struck me that the notion of joining was an inaccurate and misleading metaphor. First, as Maturana points out (Mendez, Coddou, & Maturana, 1988), there is no single family to join. Each person has his or her own family, in which he or she plays a pivotal role and assigns supporting roles to the other members of the household. Thus, a therapist who sits in a room with, let's say, a mother, father, son, and daughter, is seeing four families, not one. Although there may be overlap in their concerns, each brings to the table a unique set of problems, beliefs, and vested interests. The son's problem with failing math is related to, but not identical with, the mother's problem with having a son who is failing math. In any event, it is not possible for a therapist to talk to or join "the family" for the additional reasons that families (a) do not have ears, and (b) have no admissions office. At best, the concept of joining a family is a metaphor, and not a very useful one.

An additional problem with the joining metaphor is that therapists often seem to want to have it both ways. They want to position themselves outside the family to avoid being sucked in to its traditions and rules. At the same time, they wanted to be inside in order to exert influence. Needless to say, one cannot be a part of and not a part of a system at the same time. In fact, from the perspective of structural determination, the outside position is more useful because it helps the therapist generate *orthogonal interaction*—that is, to interact with components of the system in novel ways.

What family therapists can, in fact, do is create one or more new organizations, comprised of themselves and one or more family members. These new organizations can be structured to have the flexibility to pick and choose among the characteristics of the existing systems. This may be a more theoretically useful way to describe what therapists mean when they say they are joining a family.

Maturana in the Original

As I thought about this and other clinical operations, I realized how much I had been influenced by Dell's writings. I decided that it was time to read Maturana's work in the original. After a few halting attempts, I pondered whether his prose might be easier to understand in Spanish. Thus, I enlisted the help of a bilingual friend, asking him to read something by Maturana in both the Spanish and English versions. Unfortunately, he reported, both versions were equally dense. Of course, I now understand that Maturana's headache-inducing circular locutions were not attributable to his poor grasp of English but were necessitated by the mind-boggling task of needing to use language to explain language.

Along with a group of other therapists, I subsequently had the good fortune of spending three inspiring days listening to Maturana discuss his work (courtesy of Paul Dell and Eastern Virginia University). I also attended virtually all of his presentations in Philadelphia (most under the auspices of the American Society of Cybernetics and the Philadelphia Society for Adolescent Psychiatry). For some reason, Maturana was easier to understand from his verbal presentations than from his writings.

After these experiences, I felt conversant enough with his perspective to write something about structural determination for an audience of clinicians. The result was a *Family Therapy Networker* article titled "The World According to Humberto Maturana" (Efran & Lukens, 1985). Please do not blame me or my co-author for that title—we would never have dared to suggest something that grandiloquent. It was Richard Simon, the editor, who affixed that sweeping label to our rather modest offering.

Our goal was to provide an accurate version of Maturana's theory, geared to practitioners. To achieve that purpose, we relied on a series of metaphors, from Walt Disney World to Alcoholics Anonymous. Did Maturana approve of the result? I have no idea. He never complained about the piece, but he was often too polite to criticize the work of others—and, come to think of it, I was too polite to ask if he had read it.

Individuals Redux

A funny thing happened on the way to bringing Maturana's worldview to the attention of family therapists. Just as we seemed to be making progress, family therapy itself suddenly and unexpectedly imploded, more or less disappearing from view, practically overnight. In 2001, *The Family Therapy Networker*—the field's most widely read periodical—was forced to change its name to *The Psychotherapy Networker* in order to avoid losing readership. At about the same time, the world-renowned Philadelphia Child Guidance Clinic—up until that point a beacon for family therapists— began losing steam and ultimately closed.

Whereas family therapy was once the new kid on the block, it was now upstaged by newer, shinier treatment regimes, such as EMDR (eye-movement desensitization and reprocessing), as well as emotion-focused therapy and an assortment of mindfulness, somatic, trauma-focused, and attachment-based approaches. These returned the focus to the individual or, in some instances, the couple. It is as if the rugged individualism giant, which had been snoozing, reawakened, saw that it was being usurped, and quickly reasserted its hegemony. In a curious twist, Richard Schwartz (1997) began popularizing his internal systems model in which he put the system inside the individual's psyche, claiming that each person had subpersonalities or parts that represented the voices of family members. Personally, I find this a gimmicky approach that I consider a significant step backwards.

Work vs. Social Relations

Although family workers were among the first to take note of Maturana's work, his wisdom can be of equal value to those who work with individuals or couples. For instance, it can help unravel what was once a hot debate about whether clients and therapists were (or should be) co-equal partners or whether therapy had to (or should) involve some sort of power differential. Many who favored a form of radical equality recommended that therapists totally abdicate the role of expert and cease any activities that involved diagnosing, interpreting, or advising. They were advised to adopt a so-called not-knowing stance (Goolishian & Anderson, 1987), acknowledging a client's presumed knowledge of their own situation. In my view, this never actually eliminated hierarchy; it mainly disguised it. After all, it was still the therapist—not the client—who set the fee, determined session length, provided the locale, scheduled meetings, and so on. Ironically, clients were not even given a choice about whether or not they wanted their therapist to be a mental health expert or a not-knowing co-equal. It was Stuart Golann (1988, p. 63) who labeled this experiment in therapeutic democracy a "disingenuous or misguided form of therapist powerlessness." He argued that it confused clients more than it helped them. It certainly did little to enhance or sustain family therapy's popularity. Moreover, although clients may indeed be experts on certain aspects of their experience, therapists can bring a different, and very useful, perspective to the endeavor. Patients may be an excellent source of information about their symptoms, but the medical doctor's knowledge of anatomy, illness, and treatments is required to assure a favorable outcome.

From my perspective, the debate about therapeutic hierarchy could have been shortened or bypassed entirely if therapists had paid attention to Maturana's distinction between social and work relations: Unlike social relations, work relations involve a product and imply hierarchy. Therapy is clearly a work relation, even if the products involved, such as improved mental health, increased life satisfaction, or avoiding divorce, are sometimes vaguely defined. Therefore, although it is certainly important that clients and therapists form a working partnership, they play different roles and have different stakes in the enterprise. In other words, some form of hierarchy is required and, generally speaking, is both useful and acceptable.

Insight vs. Change

Maturana's theory also helps us understand the relationship between insight and change. Long ago, psychiatrist Thomas Szasz wrote that the "early Freudians believed that insight 'cured'" (1973, p. 80). They eventually gave up on that idea, but—as Szasz sardonically notes—they never bothered to offer anything to take its place. For Maturana, explanation (which would include insight) consists of the reformulation of a phenomenon in an alternative cognitive domain. Explanations neither take the place of, nor necessarily modify, the phenomena being explained. Thus, after explaining lightning in terms of electron theory, you can still be in awe of a stormy sky or manage to get yourself electrocuted if you aren't careful.

In structural determination, explanations are considered secondary, not primary. From this point of view, because explanations and results are in different domains, it is clear that clients can change (or remain the same) with or without insight. Parenthetically, we should add that it is typical for many people, having achieved the results they were seeking, to lose interest in explanations. Moreover, some clients who have not achieved satisfactory results treat explanations as a kind of consolation prize. For instance, they admit that they still procrastinate as much as ever but add that they at least now know why.

Conservatism

Maturana asserts that life is conservative. If that is so, why do so many clients tell therapists that they want to change or want to be different? The truth is that they do not mean it! They do not want to be different. What they want is to maintain (or regain) some quality, possession, or circumstance that they perceive themselves to be losing or having lost. In other words, although they talk the language of change, their motives are conversative.

Consider, for instance, people who claim that they want to stop smoking. In reality, what they want is to continue smoking but without running the risk of getting cancer or suffering from additional spousal criticism. Their problem is that they have not found a way to maintain the benefits of smoking without incurring the unacceptably high costs that come with it. The therapist's job is to listen to the client's "change talk" and then assess exactly what he or she would like to conserve (or, if necessary, resurrect) and which costs he or she feels a need to minimize (Mendez et al., 1988). Armed with that information, the therapist can assist the client in devising the best cost-benefit solutions.

Aristotle's Curse

It was personality theorist George Kelly (1969) who noted long ago that the traditional Aristotelian distinctions between thoughts, feelings, and actions "confuse everything and clarify nothing" (p. 91). Thus, I am grateful that structural determination provides a more coherent, biologically grounded schema for discussing and managing these core human experiences. In a nutshell, language is best viewed as a form of communal choreography. Words, as Maturana cautions, should not be deprecated. They can start wars, end careers, ruin relationships, and provoke suicides. On the other hand, the term emotion is best used to refer to the bodily settings that support (or inhibit) various classes of action. Note that, in this model, thoughts and emotions are not construed as being in competition or in opposition to one another. Moreover, emotion does not just refer to emergency flight-fight reactions or extreme experiences of joy and grief. On the contrary, shifting bodily calibrations are a continuous aspect of our personhood, and we require the appropriate hormonal and skeletal adjustments in order to sleep, eat, reason, make love, and so on. When these settings and the activities required are out of synch, we experience the vexation that Maturana labelled an *emotional contradiction*. For instance, a person is still steaming mad but socially obligated to smile and act as if everything is fine. A bit of emotional contradiction is normal, but too much of it drives people to seek therapy.

Hopefully, the vocabulary of structural determination will finally result in wiping out the remnants of the culture's primitive steam-kettle thinking—the biologically ludicrous notion that emotions are stored entities that have to be released before they boil over or cause various forms of psychic mischief. Too many clients and therapists still conceptualize therapy as being primarily about this release of feelings—a place to get things off your chest. From my viewpoint, therapy is about discovering more satisfying living patterns, not engaging in cathartic rituals.

Cures, Quotes, and Questions

The more I think about Maturana's work, the more I recognize the multitude of ways in which his thinking has shaped how I practice and teach (Efran, Lukens, & Lukens, 1990). For example, I now use the term *cure* to mean that the client's questions have been answered (at least for the moment). Unanswered questions gnaw at us. When they have been answered, they no longer dominate the conversation we have with ourselves. In many cases, clients are stymied because the forms of their questions do not permit useful answers. For instance, a client wants to know how to increase his self-esteem so that he will be more motivated to seek a job. Of course, if he succeeded in getting a job, he might report having increased self-esteem! (One might say that self-esteem is what you get when you don't need it anymore.) Thus, the way the person conceptualizes his problem keeps him stuck in a counterproductive loop.

As Maturana notes, the form of the question dictates the form of the answer. Therefore, to escape the quagmire in which a person is trapped, he or she must change "the nature of the question, to embrace a broader context" (Maturana & Varela, 1987, p. 135). Einstein voiced a similar sentiment, often saying that "you can never solve a problem on the level on which it was created."

On a lighter note, I am amused by how often I have had good results simply by quoting bits of Maturana-ese. For instance, I have advised several dominated spouses that authority is always by concession. I have reminded couples that their conversation of "accusations, characterizations, and recriminations" (Mendez, et al., 1988, p. 158) is unlikely to result in problem-solving. Moreover, problem-solving conversations are in a different (conversational) domain. You cannot get from one to the other—you have to make a choice and take a leap: Do you want to continue making your partner wrong, or are you more interested in solving the problem at hand?

In some settings, Maturana's odd phrasings can be a hindrance. However, in therapy, they can be useful because they grab attention. The familiar puts us to sleep; the novel wakes us up. Ironically, depending on a client's level of sophistication, I have gotten lots of mileage by exploring the client's understanding of four of the most cybernetically troublesome concepts: control, change, cause, and choice—notions they think they understand, but don't.

Life As a Natural Drift

In my work and my life, I have also made good use of two broad philosophical premises that align with Maturana's perspective. First is the proposition that life is never personal. In other words, nature doesn't care. It doesn't root for either my dogwood tree or the fungus that has attacked it. As someone once said, "nature works as steadily in rust as in rose petals." Not even the encounters that we regard as the most intimate and personal, such as the experiences of being loved and hated, are personal in the way we typically presume. No matter how intense, they simply remain a function of structural fit. It is important to keep in mind that the macho driver who cuts you off on the highway is just doing driving his way. He does not know who you are and is not trying to ruin your day. Moreover, the person who falls in love with you would fall in love with anyone who happened to trigger his or her personhood in the same way you do. It is much easier to navigate life's struggles when you give up the notion that things are personal and directed at us in a vicious way. The problems you experience take place where you are and you are therefore involved in (i.e., implicated in) their solution. However, they do not represent a vast conspiracy against you, any more than the soda machine that fails to dispense a soda when you push the appropriate button is seeking revenge against you. You just happen to be the next one in line.

The second key principle is that life is never lived in the hypothetical. As they say, "This is it." There is only now—the past and the future are language inventions. A major function of the past is to justify the present, and it fulfills that function admirably. However, the past does not extend into or dictate the future. To say you have a habit because of your past is to acknowledge that you did the same thing on Monday, Tuesday, Wednesday, and Thursday. You are still entitled to try something different on Friday. Moreover, existence is a natural drift to which—in language—we add and exchange our endlessly fabulous stories.

References

Bateson, G. (1979). *Mind and nature: A necessary unity.* E. P. Dutton.
Dell, P. F. (1981). Paradox redux. *Journal of Marital and Family Therapy, 7,* 127–134.
Dell, P. F. (1982a). Family theory and the epistemology of Humberto Maturana. *The Family Therapy Networker, 6*(4), 29, 39–41.
Dell, P. F. (1982b). Beyond homeostasis: Toward a concept of coherence. *Family Process, 21,* 21–41.
Dell, P. F. (1985). Understanding Bateson and Maturana: Toward a biological foundation for the social sciences. *Journal of Marital and Family Therapy, 11,* 1–20.
Efran, J. S., & Lukens, M. D. (1985). The world according to Humberto Maturana. *The Family Therapy Networker, 9*(3), 23–25, 27–28, 72–75.
Efran, J. S., Lukens, M. D., & Lukens, R. J. (1990). *Language, structure, and change: Frameworks of meaning in psychotherapy.* W. W. Norton.
Golann, S. (1988). On second-order family therapy. *Family Process, 27,* 51–65.
Goolishian, H., & Anderson, H. (1987). Language systems and therapy: An evolving idea. *Psychotherapy: Theory, Research, Practice, Training, 24*(3S), 529–538. https://doi.org/10.1037/h0085750
Johnson, W. (1946). *People in quandaries: The semantics of personal adjustment.* Harper & Row.
Kelly, G. A. (1969). *Clinical psychology and personality: The selected papers of George Kelly* (B. Maher, Ed.). Wiley.
Maturana, H. R., & Varela, F. J. (1987). *The tree of knowledge: The biological roots of human understanding.* New Science Library/Shambala.
Mendez, C. L., F. Coddou, & Maturana, H. R. (1988). The bringing forth of pathology: An essay to be read aloud by two. *Irish Journal of Psychology, 9,* 144–172.
Schwartz, R. C. (1997). *Internal family systems therapy.* The Guilford Press.
Selvini Palazzoli, M., Cirillo, S., Selvini, M, & Sorrentino, A. M. (1989). *Family games: General models of psychotic processes in the family.* W. W. Norton.
Simon, R. (1985). A frog's eye view of the world. *The Family Therapy Networker, 9*(3), 32–37, 41–43.
Spencer-Brown, G. (1972). *Laws of form.* Julian Press.
Szasz. T. (1973). *The second sin.* Anchor Press/Doubleday.
Varela, F. J. (1979). *Principles of biological autonomy.* Elsevier/North Holland.
Watzlawick, P., Beavin Bavelas, J., & Jackson, D. D. (1967). *Pragmatics of human communication.* W. W. Norton.
Watzlawick, P., Weakland, J. H., & Fisch, R. (1974). *Change: Principles of problem formation and problem resolution.* W. W. Norton.

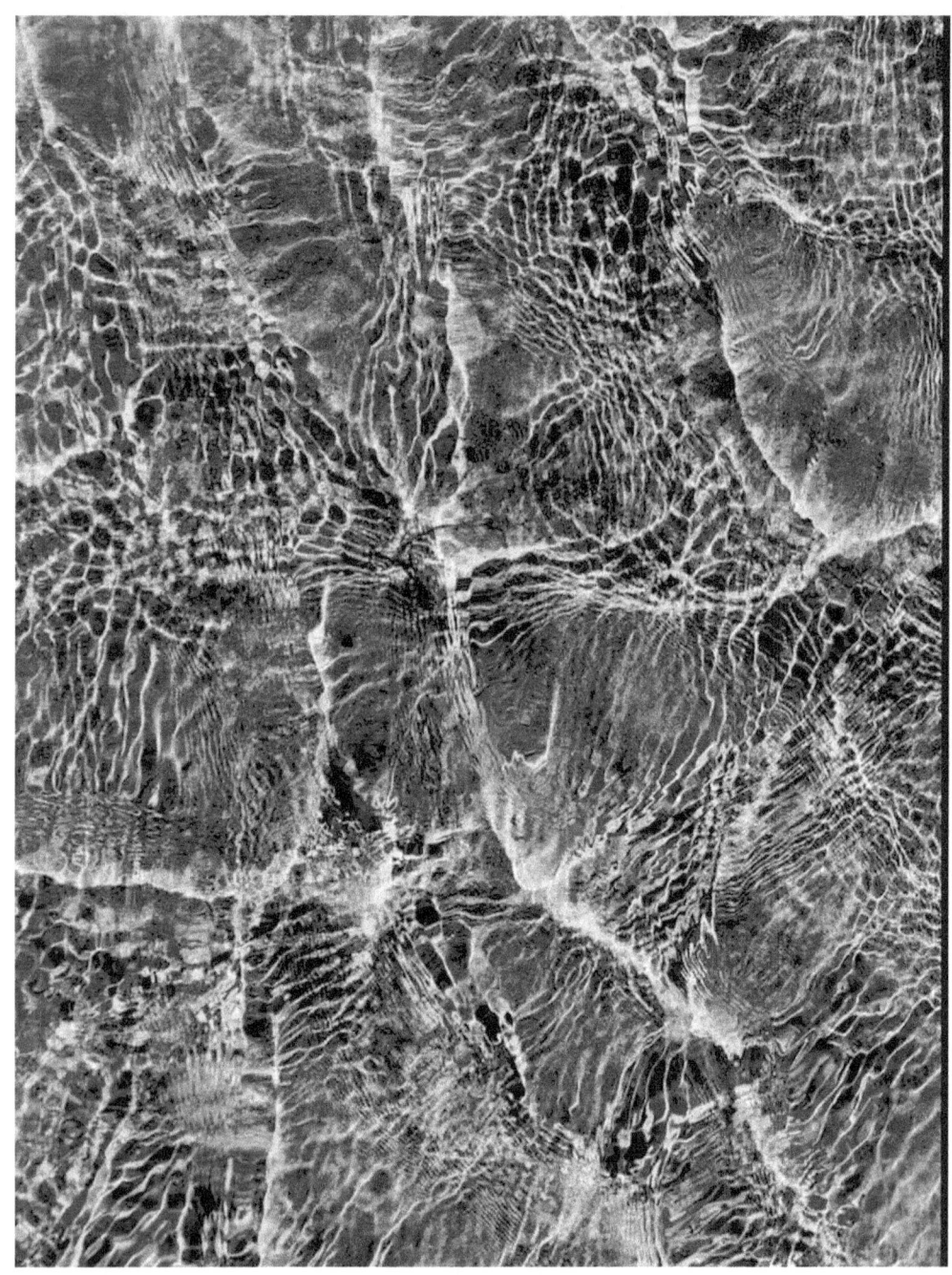

Bunnell, P. (2019). *Scintillating Network*. Photograph.

Reflections on Core Ideas of Humberto Maturana in Relation to The World Café

Juanita Brown[1] and Amy Lenzo[2]

This article illuminates Humberto Maturana's seminal ideas on the ontology of conversing embedded in the cultural-biological foundations of human existence and our co-evolution as loving beings through the relational spaces and networks of conversation in which we participate.

It illuminates the powerful role of Maturana's scientific insights on the global evolution of the World Café, which uses simple design principles to cultivate the embodiment of Maturana's concept of love—the legitimate co-arising of the other in the relational space between us—in the flow of the dialogic doings of that Café. World Café conversations reflect a simple design for emergence that enables both the relational/emotional space and the conceptual/ideational space to be braided together simultaneously, with the capacity to harvest and conserve what has held the life in that Café and beyond.

Amy Lenzo, Global Coordinator of the World Café and online pioneer adds her insights regarding how Maturana's ideas can create life affirming networks of conversation in the online world. The article also references the pioneering work of Flavio Mesquita, in this issue (pp. 45–53) who used Maturana's thinking, as well as that of Paulo Freire, the Brazilian educator, coupled with whole systems design in the implementation of Generation of Peace, the large scale educational change project in the state of Ceará, Brazil. The World Café served as the core process for personal, interpersonal, and institutional evolution in this five-year project encompassing a sprawling educational system of 700 schools, 16,000 teachers, and 400,000 students.

Keywords: Maturana, World Café, large-scale change, dialogue, conversation

---

In the summer of 2008, I, Juanita, as part of my collaboration as a research affiliate with the MIT Organizational Learning Center, spent a week with Humberto Maturana and Ximena Davila from the Matriztica Institute, along with Peter Senge and other MIT colleagues.

It was a truly inspirational and profound period of learning and discovery about the cultural-biological foundations of human existence and our co-evolution as loving beings through the relational spaces and networks of conversation in which we participate. In our informal conversations together, I could sense Humberto's deep understanding of the human condition and his hope that, somehow, his scientific insights could make a difference to our common future on this fragile planet.

As a non-scientist and organizational practitioner who had, at that time, spent more than 30 years working with global leaders, I was deeply touched by Maturana's insights as an evolutionary biologist as well his hopes for the use of his work on behalf of life-affirming futures.

Humberto's insights, as I understood them, have had a profound influence on the evolution of the World Café, an innovative approach to large scale dialogue and

---

1. Co-Founder, The World Café. Email: Juanita@theworldcafe.com
2. Global Coordinator, The World Café Community Foundation. Email: Amy@theworldcafe.com

collaborative engagement around critical issues. The World Café pattern for creating living networks of conversation was born in our living room with a multi-disciplinary group of intellectual capital pioneers in 1995.

From that initial group of 24 leaders in our living room, the World Café community of inquiry and practice has now spread to multi-sector and multi-stakeholder settings in organizations and communities on six continents. My book, co-authored with David Isaacs (2005), *The World Café: Shaping Our Futures Through Conversations That Matter,* has been translated into 17 languages, which has expanded its reach to many new international constituencies.

Later in these reflections, I'll introduce Amy Lenzo, World Café Global Coordinator and an international pioneer in online dialogue and engagement. She will add her unique perspective on the ways she has translated Maturana's insights and perspectives and integrated them into World Café conversations held on-line with participants across the globe.

The case study in this volume by Brazilian World Café pioneer, Flavio Mesquita, is a powerful example of using Maturana's thinking, as well as that of Paulo Freire, the Brazilian educator, coupled with whole systems design in the implementation of Generation of Peace, the large scale educational change project in the state of Ceará, Brazil. The World Café served as the core process for personal, interpersonal, and institutional evolution in this five year project encompassing a sprawling educational system of 700 schools, 16,000 teachers, and 400,000 students (Mesquita da Silva, 2017; see pp. 45–53 of this issue).

The following informal reflections are based on personal notes from my time with Humberto at MIT as well as my earlier research for our book on the World Café. I offer apologies in advance for any misinterpretations that I might have had during the time I spent with Humberto and his colleagues from the Matriztica Institute. His ideas are quite precise and I may not have interpreted them as fully as Maturana intended. Yet, our interpretations of his pathfinding work in the real-world settings in which we practice have had a powerful influence on the trajectory of the World Café and its global influence.

The ideas below, taken from my notes at the seminar, have a direct relationship to our work with the World Café. They provide a taste and flavor of our time together with Humberto at this week-long gathering with us on the cultural biological matrix of human existence.[3] I'll then share a description of how I see the World Café embodying some of these powerful scientific insights about the origins and conservation of our very humanness. In our time together Maturana offered that:

- As humans we are born in the trust of loving and in being loved, within an ecology of the natural world and within the larger living cosmos. Love is the legitimate co-arising of the other in the relational space between us. Love exists in

---

3. See Maturana & Verden Zöller (1996. 2008) and Maturana (1989) for a deeper investigation into these ideas.

the domain of relational behaviors through which the other arises as a legitimate other in coexistence with oneself.
- What we understand as humanness are relations conserved in love over many generations of our co-existence.
- Humanness is not a genetic mutation. It is a manner of living where there is pleasure in each other's company, sharing food, nearness, caressing and tenderness—nor is the capacity for language a genetic mutation—it is an evolutionary drift emerging from the intimacy of human community and the coordination of actions in language together.
- It is in the intimate community where humanness arises as a network of conversations that is conserved over generations as a lineage through the raising of children over hundreds of generations in manners of living that are conserved in that lineage. Humanness did not arise in competition, struggle etc. It arose in intimate family/community co-existence.
- We live in the braiding of emotions and languaging in our manner of living together. In this coordination through language, certain consensus or agreements appear as "reality" and the objects we understand as "real" appear.
- Words are not trivial—words are the nodes or elements of networks of conversation. Language is the coordination of doings, not a symbolic act as we commonly understand it. Our languaging distinguishes a way of inhabiting a human community and culture.
- As human beings we find ourselves living in communities in recursive coordination of doings, generating different worlds and realities as different manners of living together in networks of conversation.
- A person who reflects creates new worlds. All distinctions are made by an observer. Our capacity for reflection in language is one essence of our humanness. We are human beings that emerged with the capacity to reflect in language and conversations and in that we generate worlds. With one word I can follow one path and with another a different path. Our languaging distinguishes a way of inhabiting a human community and culture.
- A key question is: How, in a human system, living what it is living, does pain and suffering arise from that manner of living? Culture is a network of conversations that can both generate and conserve states and manners of living and co-existence, even painful ones.
- We have a choice and can be intentional about what we want to conserve in our manner of living and what we don't. Everything changes and evolves around what we want to conserve. Do we want to conserve our essential nature as loving beings? Homo sapiens amans?
- The Matriztica Institute invites reflection on our fundamental nature as biological-cultural loving beings who arise in language and who live in conversation as our manner of living (our doings) in conversation. Organizations, for example, are not the organizational chart, but rather arise in the living network of conversations that conserves certain manners of living together.

- What is intelligence? It is the behavioral adaptability/plasticity that we show in our manner of living in co-existence with a changing environment.
- All cultural change, for example, is a change in the network of conversations and the manner of living that arises in it. Language and conversations are doings that lie at the heart of our capacity to intentionally bring forth worlds that are life-affirming and ethical. Everything changes around what we want to conserve.

What's the World Café and What's Maturana Got to Do With It?

As mentioned earlier, the World Café is a collaborative dialogue process designed to create living networks of conversation that connect intimacy and scale in unique ways. World Café conversations embody a simple *design for emergence*, based on core design principles, that enable both the relational/emotional field and the conceptual/ideational field to be braided together simultaneously. World Café dialogues may include as few as 12 people but makes its unique contribution with large groups that may include hundreds of participants.

Figure 1: Large Group Café

In a World Café experience, intimate table groups of 4-5 people engage in deep collective listening and reflective conversation, where each member is seen, heard and honored for their unique contribution. Each table has a paper tablecloth on which the members collectively draw, write, and visually connect ideas as well as create images that strike the members as significant around the question(s) they are exploring. Participants are encouraged to *listen into the center* to discover insights that are emergent from the web of dialogic interaction among them.

Figure 2: Single Café Table With Drawings

This creates a relational space and dynamic in which what is conserved from round to round is what holds the most life from each conversation. These fertile seeds in the middle are what is conserved and taken to be offered by members as they continue into their next table round. As the conversation unfolds through these multiple cross-pollinating rounds of respectful dialogue in an intimate relational space (the small table circles), a living network of conversation emerges. What is conserved and spread through consecutive rounds of recursive conversation becomes part of the embodied experience of that living network of conversations.

Collaborative intelligence, generative possibilities, and the relational field begin to arise in tandem as more and more people, in the intimacy and caring of the small table groups, are truly seen and heard in the evolving doings of that Café's network of conversations. A World Café dialogue is, in a sense, a small lived experiment in the cultural biology of love and our fundamental humanness with each other, at increasing levels of scale.

Maturana shared in our time together at MIT that love exists in the domain of relational behaviors through which the other arises as a legitimate other in relation to oneself. At its best a World Café experience nurtures a relational space where each member arises as a legitimate presence with the other members simply for being the unique person that they are and for the contribution they are offering throughout the network of conversations as they co-evolve.

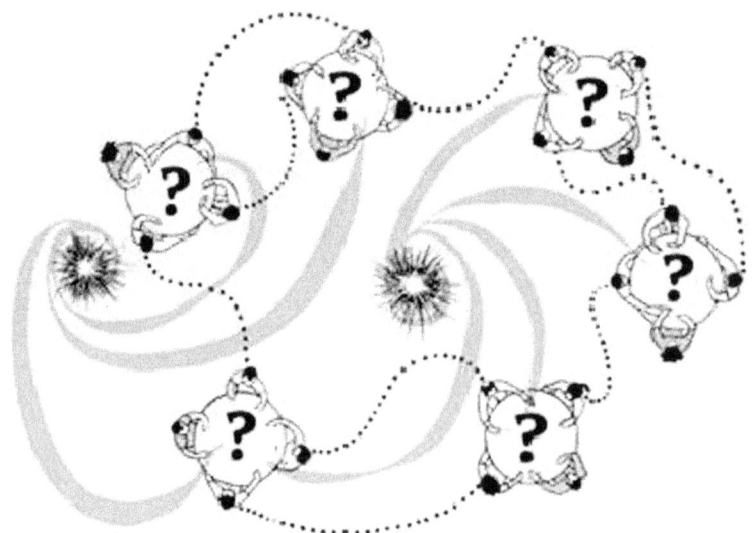

Figure 3: Living Network of Conversations

The whole group reflection and the often-used visual harvesting at the end of each Café experience, makes visible to the whole group what is being conserved (patterns, insights, deeper questions, possible coordinated actions). This visual harvest, in addition to what individual members may be taking personally, holds the life (and the living) that has emerged through the recursive network of conversations embodied in that particular Café.

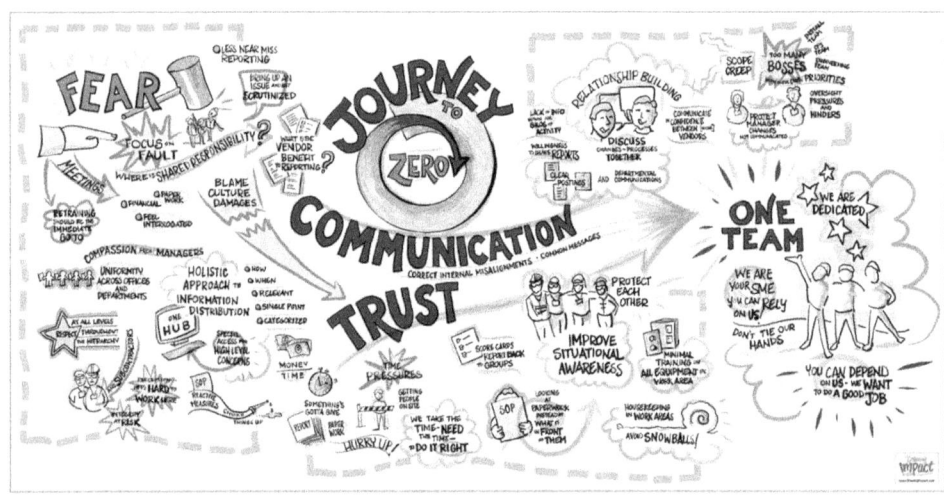

Figure 4: Visual Café Harvest

Opening Spaces for Conversation and Human-ness in the Online World

My colleague Amy Lenzo, an online pioneer, brought the World Café and its design principles into the online environment over a decade ago. In this era where we are living so much of our lives online, I asked her to share how she is using insights from Humberto Maturana in her own work, which she shares in the following reflections:

My passion in using the World Café and other participatory approaches online is to open spaces for conversations that enliven members to co-inspire new worlds together. Maturana's definition of love as the relational and empathic recognition of the legitimate co-existent other is also experienced online, only slightly differently than it is on site. Some of the connective tissue of relationship that Maturana talks about is embedded in the physicality of lived experience. But our biological make-up, our humanness, is the same no matter where we are, or how we gather.

It is this humanness, this innate ability—no, drive—to connect with each other in intimate and meaningful ways through conversation, that distinguishes World Café conversation from casual everyday interaction.

The importance of language as the foundation of our understanding of reality is crucial to World Café conversations. Language, to paraphrase Maturana, literally creates the worlds we live in. Our words matter; our questions matter; our language matters.

When we are online, language is narrowed. And yet, through that narrowing it is also deepened.

By language here, we are not only referring to the words we choose to convey our meaning, which are of enormous importance, but also to communication, language, received through the other senses we have access to online: visual, in our capacity to convey meaning through shared images, through our gestures and facial expressions, through background images and personal presentation; and, auditory, in the sense of tone, pace, volume, rhythm. The use of music and silence.

The intuitive and empathetic senses—in our ability to engage with each other, in small groups, and across distance. To lean into, and experience what it is to be a human being in this strange new world—to experience each other with our feet on the ground in different physical locations, literally all over the globe, at the same time as we occupy a shared space together in real time, online, across time and distance. And finally, the sense of imagination, not in the sense of making something up that isn't real, but in giving image to something that does exist but would not otherwise be visible.

This combination of sensory perception expands the language we use to create the online environment and transforms what we can imagine being possible as human beings alive on this planet today.

The World Café process reliably creates a relational field that enables us to see each other's contributions as legitimate, even when they are different from our own. In this way, World Café conversation has embedded Maturana's insights about love to an extraordinary degree. And we have found that this relational field is equally present online.

Martin Buber, in *I and Thou* (1984), says all real living is meeting. Expanding the scope of how many people have been touched by the shift in perspective that the World Café engenders, World Café has also been a global pioneer in bringing this level of meeting online, where the field is unbounded by time and geographic space and the possibilities are still wide open.

What I have learned in over a decade of bringing the World Café and other participatory practices into an online environment, is that this work is no less real and powerful online than it is when experienced in the flesh; working online, we can also enable and evoke the I-and-Thou relationship that Buber (1984) speaks of. A recognition and evocation of this relationship is embodied in the World Café process, and it is this level of recognition of each other as *legitimate other* that Maturana speaks about as love. And love is the ultimate ground for our work, both online and onsite.

Closing Reflections by Juanita

On the last day of our MIT group's time together with Humberto and his colleagues, he shared a personal aspiration and an invitation to those present. He shared that:

> As a co-inspirator, I can be intentional about the nature of the conversations I introduce into the conversational network that is the organization or the culture I am part of. This is serious, responsible, daring, and playful work! How I open spaces of conversation is of the utmost importance to our capacity to co-inspire worlds we choose to live in.

It has now been 26 years since the birth of the World Café. It continues to co-evolve in response to the needs and challenges of our times. In this era of widespread divisiveness and strife, Maturana's pioneering scientific insights about the biology of love, along with the power of conversation as a co-evolutionary force are now more important than ever. He makes clear that only love—the capacity to engage others as legitimate beings across our differences—will nurture the co-intelligence needed for our species to continue to evolve in generative, life enhancing ways.

It is our fondest hope is that the global World Café community continues to honor Humberto Maturana's legacy by continuing to play a role as co-inspirators who are increasingly adept at opening brave spaces for generative conversations across the globe, both online and onsite, that point toward life-affirming futures worthy of our very best efforts.

Acknowledgment

Portions of this article appeared as a World Café blog (8/18/08) based on the August 2008 MIT Organizational Learning Center's week-long gathering with Humberto

Maturana & the Matriztica Institute on the ontology of conversing as part of the origin of humanness and the biology of love.

References

Brown, J., & Isaacs, D. (2005). *The World Café: Shaping our futures through conversations that matter.* San Francisco: Berrett Koehler.
Buber, M. (1984). *I and Thou.* New York: Scribner Classic.
Maturana, H. R. (1989). *Ontologia del Conversar* [The ontology of conversing]. *Persona y Sociedad, III*(2), 9-28.
Maturana, H. R. & Verden Zoller, G. (1996). The biology of love Munich: Ernst Reinhardt.
Maturana, H. R., & Verden Zöller, G. (2008). *The origin of humanness in the biology of love.* Charlottesville, VA: Imprint Academic.
Mesquita da Silva, F. (2017). *Generation of peace dialogues: How the World Café approach to community understanding led to cultures of peace.* Dissertation, Fielding Graduate University. ProQuest Dissertations Publishing, 10601890.

Bunnell, P. (1999). *Maturana as Joyful.* Photograph.

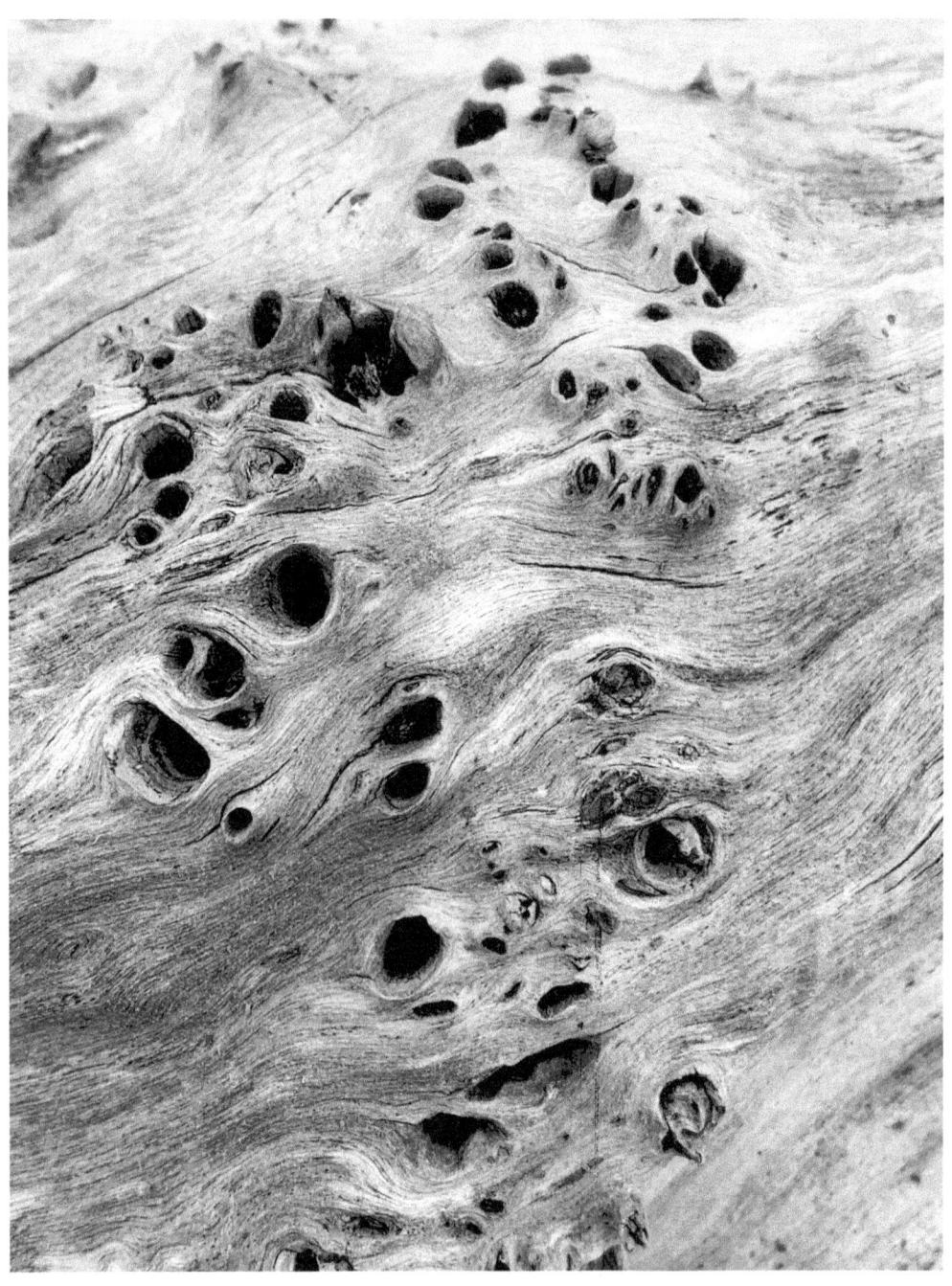

Bunnell, P. (2019). *Branch Points*. Photograph.

Designing for Emergence:
Creating Living Networks of Conversation Grounded in Love

Flavio Mesquita da Silva[1]

The seminal work that Humberto Maturana brought forth about the biology of love and the ontology of conversing sheds light on the path to understanding our humanness. Accordingly, he proposes that our co-existence requires that our interactions be grounded in love, which entails that love is the dynamic condition for the creation of sustainable living networks of conversations. This article reflects on Maturana's foundational contributions to the design framework of a large-scale peace project in Brazil, where the World Café, a participatory approach to large-group collaborative dialogue, provided a living laboratory for Maturana's core ideas along with those of Paulo Freire, and others.

Keywords: Maturana, whole systems design, love, dialogue, peace, World Café, critical consciousness

---

This reflective article is an account of a large-scale peace initiative implemented in Brazil, the *Generation of Peace*—a collaboration between the Brazilian Government, UNESCO, and the State Department of Education of Ceará (SEDUC). I had the honor of designing and coordinating this project in the school system, over a five-year period from August 2010 to March 2016. The project scope encompassed 700 high schools in 26 school districts, with 400,000 students and 16,000 teachers.

The project design was grounded in the extensive work of Humberto Maturana, whose foundational contributions about the biology of love (1996, 2008) and the ontology of conversing (1989) helped bring forth the change theory that I applied throughout the project.

In addition, Juanita Brown and David Isaacs's pioneering contributions regarding the role of living networks of conversation and cross-pollinating diverse perspectives as catalysts for systems change were key to the whole systems design of the peace initiative. Their work with the World Café (2005), a participatory approach to large-group collaborative dialogue around critical issues that has spread to six continents, is also based on Maturana's insights and was the core process used in the Generation of Peace initiative (see Brown & Lenzo, pp. 35–43 this issue).

The increase in the rate of violence among youth in the State of Ceará was serious; especially in Fortaleza, the capital of the State. Fights, bullying, and drug trafficking on and near school grounds were widely reported. The top down system of education as well as the local school climate affected the students' academic achievement and resulted in many abandoning their studies before completion. In addition, the

---

1. Email: fdasilva@email.fielding.edu

socioeconomic situation of some regions in Ceará, particularly related to public safety, aggravated the vulnerability of certain marginalized communities.

The public outcry for immediate solutions to the growing violence and other related critical issues was intensifying. In spite of valiant efforts to respond to these multifaceted dilemmas, SEDUC's leadership was aware that the dramatic challenges it faced would require a comprehensive whole systems approach. This was the situation when I was called in to be of support to the design and implementation of what came to be called "Generation of Peace" in both of it's meanings in Portuguese: the generating of peace; and a generation of peace.

The Generation of Peace initiative, and the whole systems design that emerged, formed the core of my dissertation (Mesquita da Silva, 2017) at Fielding Graduate University (*Generation of Peace Dialogues: How the World Café Approach to Community Understanding Led to Cultures of Peace*).

At the heart of the project was a concept we called peace for peace's sake. Instead of focusing on fixing or stamping out the violence as a dramatic and dangerous problem, we chose to focus on the generation of peace as an opportunity and a possibility that everyone could live in their own lives and spheres of influence. Our methodologies were designed with this appreciative approach in mind.

These approaches focused on co-evolving more generative ideas about the nature of schooling itself as well as the systems where schooling occurred—the family, the school itself, and the greater community, all of which needed to play both direct and indirect roles in creating more life-affirming outcomes.

It implied more than school reforms, whose framework historically had focused mainly on the betterment of schooling in economically disadvantaged districts. It entailed a transformative process that also addressed other subtle forms of poverty that are present in any Brazilian community, including modern, rich, and urban ones.

Design Framework: Key Influences

As mentioned earlier, the rest of this article will focus on how key insights of Humberto Maturana along with the practical application of the World Café as a core process were embodied in the Generation of Peace initiative in Brazil. However, our overall design framework was also influenced by the remarkable work of Paulo Freire (1998, 2011), on critical consciousness and dialogue, and Gregory Bateson (1972, 1980) on an ecology of ideas and meta-communication, among others.

Maturana suggests that the evolution of culture itself is embodied in "a flow of conversing within a particular network of languaging and emotioning" (Maturana, 1989, p. 4). That being the case, generating a culture of peace in a huge school system like that of Ceará could not thrive without serious initial work on fostering a culture of dialogue and mutually respectful conversation across the multiple social systems and diverse perspectives encompassed by the project.

Assuming that love, according to Maturana (2002a, p. 185), "is the source of socialization, not a result of it," I affirm that love is a natural yearning that allows for being together as long as the conditions are favorable for the interactions that recognize, value and sustain one another. Further, Maturana asserts that in interpersonal relationships, "love is the dynamic condition of acceptance, of its coexistence with the other (or others)" (p. 184).

Therefore, to love is to open a space of interactions with the other, in which their presence is legitimate, no requirements. In contrast, rejection constitutes the space for behaviors that deny the other as a legitimate other in coexistence (Maturana, 2020). Although tolerance does not equal love, living our interpersonal relationships accepting difference, even if temporarily, rather than rejecting one another can be a sign of hope to create opportunities for dialogue consistent with the atmosphere generated in the World Café.

Freire's *Pedagogy of the Oppressed* (2011) centers on people's capacity to understand their social reality through reflection and action, which he called *conscientization.* Conscientization implies each person's progressively clearer perception of their reality, and their own perspectives, "as interacting fragments of the whole" (p. 104).

Each World Café dialogue, with its evolving network of conversations emphasizes the awareness that each person's perspectives are a unique contribution to the unfolding whole as the group explores questions that matter to the life of that Café.

The dialogical nature of the project's design needed to take into account varied institutional structures and processes, as well as the cultural and human aspects and their numerous relations within the overall educational ecosystem. The design aimed at communicating with all parts of the educational system as a whole. We did this by creating architectures of engagement that fostered networks of conversation that could embody relevant, respectful (and fun!) conversations needed in order to build mutual respect among all participants in varied sectors.

Maturana's core principle of love being embodied by cultivating the capacity of each participant and their perspective as a legitimate contributor in the Peace Café's growing network of conversations was key to our thinking.

Throughout the project the design of the Peace Cafés as they came to be called, helped to create conditions that enabled a growing mutual acceptance and legitimacy of each participant whatever age, department, or function was represented, through which all could be inspired by and nourished with an ideal of peace as reflected in their own part of the larger system.

The early Cafés involved the communities of two macro regions (nine school districts) totaling more than 3,000 people (students, teachers, school managers, governmental and non-governmental representatives, and community leaderships). The Cafés ranged from 30 to 850 participants, who helped identify the design questions as they reflected on what they wished for themselves and for their communities, including future generations. Participants shared stories and images as they explored the early, fundamental questions derived from the following: What's peace? What's peace made of? What does peace produce?

A Living Metaphor Arises

These initial Peace Cafés gave rise to a more evocative and powerful question, which in turn allowed for more clarity of the potential pathways to organizing the project both regionally and locally, as a whole system: What is the face of peace?

More than a desired vision of the future, this question offered a *generative metaphor* that allowed for co-generating images of peace in the here and now, and beyond—faces of peace specific to each school, as well as other levels of the system. These were developed through a combination of several means including posters, collage, poetry, songs, and other creative approaches to imagining the face of peace.

These heartfelt, creative representations became opportunities for further recursions of consensual coordinations of behavior in both language and imagery in an orientation that was conducive to structural changes that favored meeting both individual and collective needs to communicate, decide, plan, implement the peace process locally. It also allowed participants to exchange their experiences and images of peace with other schools regionally and across the state.

According to Maturana, Yánez, and Muñoz (2015), below, the shift from the way of asking three core questions into one even more evocative question resulted in an evolution of both the framework and the definition of these networks of conversation as *niches*, which unfolded, over time, as meta niches as they started to integrate larger forms of organizing—regionally and statewide. The work followed the implications of Maturana's observation that:

> love is the fundament of the possibility of the arising and the existence of the molecular autopoietic system and the dynamic ecological niche that makes it possible as well as the condition of possibility of the relational nature of the realization of our human existence as socially and ethically conscious living beings. We call this fundamental relation for the existence of anything, ecological love relationship. (Maturana et al., 2015, p. 633).

The whole systems approach to the case for dialogue and peace that the project entailed justified a comprehensive focus that could enable the creation of unique, sustainable niches of local and regional actions while guaranteeing the preservation of the systemic integrity of each of these niches and their possibility for learning from one another. Therefore, the term that I thought best conceptualized peace in such a broad approach was *systemic wholeness*. Systemic wholes as a concept had the power to represent each of its constitutive elements alone as well as in relationship with one another.

The experience of the Peace Cafés in many contexts with different populations and number of participants reaffirmed the necessity of having a common idea of what designing an effective World Café entails. The idea of challenging both the living (theory) and lived (experience) principles of the World Café was aimed at finding out whether these design principles would suffice in the context of macro social-architecture processes like the Generation of Peace project. The design principles, based on initial World Café research on the lived experience of key hosts around the globe are:

- Set the context
- Create hospitable space
- Explore questions that matter
- Encourage everyone's contribution
- Connect diverse perspectives
- Listen together for patterns, insights and deeper questions
- Harvest and share collective discoveries

The original principles served us well. They survived the process. However, the diversity of the situations that occurred in the course of implementing the project required that these principles be adapted according to each circumstance. This required more sophisticated design, planning, and management while remaining simple and comprehensible enough to be useful wherever it was applied. At times, a combination with other methods and techniques might be necessary as well as the use of alternative resources. In sum, the World Café as a methodology turned out to be flexible enough to encompass other approaches and means of embodying the faces of peace at all levels of the educational system.

Using a whole systems design approach, we had the opportunity to test adaptability and adjustability of the World Café approach to variables such as context, duration, format, scope, logistics, organizational climate, weather, and both cross-cultural and multigenerational aspects for more than five years. While designing, conducting, reporting, and assessing an unfolding process with thousands of co-participants, we, by necessity, had to learn to rely on some general principles that could assure the operational conditions within which the World Café could best serve the project.

The framework below might help explain how this crucible worked in supporting this statewide, long-term process of conversation, engagement and embedding the faces of peace in the school system of Ceará in many different ways.

I conceived of the project on multiple levels. A single Peace Café around the question "What is the face of peace?" created cross-pollinating links of participants moving from table to table, discovering, in Bateson's terms, the patterns that connect, or in Maturana's framework, what is conserved in the life or doings of that Café. As the question traveled to other Peace Cafes, we could begin to imagine the entire school system encompassing 700 schools, 16,000 teachers, and 400,000 students as one huge World Café, with multiple tables and constituencies interacting around the same core question in a variety of organic, emergent ways that were relevant to the stakeholders who were exploring the question.

The "What is the Face of Peace?" question is more than a question. It is a *generative metaphor* that stimulated the creative and collaborative imaging that unfolded in each Peace Café. The question was offered at some point somewhere in each specific Café. It could be at the very beginning, in the middle or at the end depending on the case. The many creative ways that the faces of peace were expressed also served to turn the participants' own living metaphors and images into means of communicating, planning and acting on what members consensually co-evolved in their conversations.

This design for emergence enabled not only the question to travel at multiple levels of the system, but also enabled what was being conserved to be shared among the students, teachers, community leaders, parents, local police dealing with school violence and drug prevention, and other stakeholders who participated in both local and regional Peace Cafés.

Four elements (or qualities) were integrated into the overall design process: *redundancy, recursion, proportion,* and *evolution* (Bateson, 1972, 1980; Maturana & Varela, 1987; Maturana, 2002a).

The redundancy of the question at the core of most Cafes, "What is the Face of Peace?" helped people make repeated connections with their own collective experiences and the growing number of next steps (i.e., possible personal as well as local school actions along with administrative/institutional policies in response to the question). In addition, there were subjective elements that helped to communicate the feeling underneath the ideas (i.e., drawings, stimulating talks, poetry, dramatizations, music, etc.) across the rounds in individual Cafés and between Cafés at the local and regional levels.

Step by step, people made sense of what was going on and progressively acquired the comprehensiveness of perspective needed to create the basis for coherent implementation of what each could influence at their own level of system.

Recursion complements the element of redundancy, since both act upon one another. While redundancy helped participants stay on track by repeating rounds of conversation within the growing network of conversations, recursion helped people, by means of reinforcement as well as cross-pollination of new ideas and insights, shift their worldviews to what emerged from each iteration, where each cycle could create successive approximations of a good fit for the needs for change and adaptation as the situation unfolded.

Proportion was also a necessary component of the design framework, in order to keep a logical flow between the rounds and the focus of each round. It supported key elements of the emergent, evolving process to be focused in the same general direction, enabling the growing of coherence without control at increasing levels of scale. This concept is a representation of the utmost definition of what conditions should be in place in the creation of a hospitable space and how these hospitable conditions should be conserved by the means of maintaining balance between the rounds of an individual Café. For example, unexpected outcomes across a Café might require some calibration of the landscape, soundscape, and so forth, so that the space remains hospitable.

Evolution is the result of an unfolding process of coherence without control—the outcome of redundancy, recursion, and proportion that resulted in an increasingly successful trajectory of the core ideas that were being conserved within the ever widening network of conversations that the project engendered.

These core ideas often journeyed from one Café to another. The ideas that were conserved or that had the most life over many Cafés found their way into both personal and institutional change over the years of the project.

As people in key roles at all levels of the system began to converse with these patterns in mind, they gradually acquired understanding of how the educational system operated, and how it could embody core ideas of peace at both the local and regional levels.

The Peace Café process also entailed acknowledging the existence and co-existence of different perceptions about the same phenomena that constituted life in the school system. The peace dialogues played a crucial role in bringing people together, despite their initial experience of different realities (Nicolescu, 2010). The focus on dialogue, reflection, and collaboration around core ideas held in common around the key question "What is the Face of Peace?" enabled people, over time, to evolve coherent decisions about what mattered to them within their own spheres of influence.

Moreover, people are more likely to recognize their multiple perceptions of reality, and to find their own and others' locus of participation in dialogic learning with one another while still being "mediated by the world," outside of the dialogue setting (Freire, 2011, p. 80). While this process can be very humbling and often chaotic, it facilitates the cross-pollination of ideas and insights that have the potential to foster the emergence of collective intelligence and wise action.

The World Café proved to be an effective system for transformative learning in multiple settings. The Peace Cafés provided the participants with many opportunities for not only acquiring new information and knowledge, but also enhancing their intellectual fluency and systemic understanding as the significance and importance of the themes that evolved in the Peace Café rounds of conversation amplified and deepened their perspectives of what might be possible, not only in individual schools but at other levels of the administration in order to bring peace alive in multiple ways.

The World Café played a decisive role in creating an environment for participants to become wiser together; it gave rise to a feeling of togetherness that fostered qualities such as tolerance, attentiveness, and empathy. And, while Peace Café participants had their reasoning and emotioning changed (Maturana & Verden Zöller, 2008) together in conversation, they not only both reflected-on-action and reflected-in-action (Schön, 1983) but also "reflected-in-interaction" (Thompson, Steier, & Ostrenko, 2014, p. 4).

Project's Potential and Its Successful Outcomes

The Generation of Peace became an exemplar, a project of the people that could offer unique opportunities for public policy that could transform the entire state educational system.

The project's potential and its successful outcomes, thoroughly described in the technical reports of SEDUC and UNESCO, as well as in the appendix of the Latin American edition of the World Café book *El World Café y la Generación de la Paz en Brasil* (Mesquita da Silva & Mesquita, 2016), show a wide spectrum of the presence of Generation of Peace in the state of Ceará at the formal completion of the project, in March 2016.

Other important achievements supported the growing development of this network of a culture of peace. This was reflected in an enhanced focus on the schools themselves in the educational system. Under this paradigm, the local schools were then intended to become the center of the system (not the periphery), as the birthplace of all key educational public policy, produced with and for the people.

The following statement gathered at the end of the final Café in March 2016, where SEDUC's leadership gathered to assess the project and plan its institutionalization in the whole educational system, translated the group's vision about the future of the institution and one of the potential benefits of this project, which could be relevant to others:

> The participants support the implementation of the culture of peace as a mission of the beings that comprise the educational system of Ceará. We need to transform the Generation of Peace into a *mark* present in everything: in the curriculum, in the management processes, in the pedagogical teaching curriculum, in the formation/training of teachers, in the educational system's projects and programs, in the collegiate bodies, in events and in the physical spaces we use. We need to embed these ideas of peace into the mission of each public servant of the system: the doorman, the cook, the teacher, the principal, coordinators, etc. (Mesquita da Silva, 2017, p. 141)

This statement was probably the most remarkable indicator of the success that we had hoped for because it encompassed the entire educational system within the scope of the project and offered very important elements for its continuation (mission statement, vision, and values). It also articulated robust arguments for creating new public policies for education. It exemplifies the reification of the concept of peace as systemic wholeness in both human and organizational systems from the perspective of the participants themselves. Capra (1996, p. 298) corroborates: "The success of the whole community depends on the success of its individual members, while the success of each member depends on the success of the community as a whole."

This Is the Boarding Gate!

The Generation of Peace Project provided immense opportunities for thousands of people to self-organize in their schools and communities, and the World Café provided a living laboratory for Maturana's core ideas along with those of Freire and others.

As we shared at the end of most Peace Cafés: This is not the baggage claim area; this is the boarding gate! With these words, participants received the incentive to fulfill the promise they made in the Cafés to make the changes they planned. It also served to set a context between Cafés, or an invisible link to next Cafés, which actually happened in most regions. This linking strategy helped the participants to connect their initial commitment to ongoing accountability while giving them a sense (a taste and a vision) of a full planning cycle (i.e., from situational diagnosis to assessment). It is the baggage, our harvest from this journey, which I hope provides insight toward the continuation of the process the project brought forth and similar initiatives as we carry our lessons learned into the future.

After convening thousands of citizens and witnessing their efforts and enthusiasm in organizing their spaces of conviviality inspired by a shared vision of peace, the legacy of the Generation of Peace Project is remarkable and might become a benchmark for public policies, not restricted to the educational sector as long as it receives the adequate support and appropriate theory of practice needed for a multifaceted long term emergent project of this type.

The gradual incorporation of the World Café into the daily practice of school system members directly and indirectly associated with SEDUC served as a form of integrating interests and as a means for self-empowerment. The awareness of this spontaneous adoption of the World Café and other practices of peace stemming from their lived experience expanded the scope of our direct attention and made the consultation team realize that there was more to learn, both for us and for the World Café itself.

Knowing that the World Café "relies on improvisation and learning-through-use and is continuously evolving" (Steier et al., 2015, p. 218), the design of each Café welcomed adaptation to unanticipated situations, which taught us how to extend its support to applications in different areas and for different purposes.

Raising individual and collective awareness and creating an organizational environment that learns was a critical task that required, in Freire's (1998) words, enthusiasm, faith, and hope. Mindful learning is a foundational element for this awareness raising to occur, which as Langer (1997, p. 4) defines it, "is the continuous creation of new categories, openness to new information, and an implicit awareness of more than one perspective."

Creating a culture of peace has a direct connection to a lifestyle and worldview, which includes the way people relate with one another on a daily basis as well as how they make decisions and choices. In addition, we can understand the participants' visions of the faces of peace as a metaphor for creating an entire culture of peace, where Maturana's idea of love as the co-arising of the legitimate other and their views, even when different from one's own, is a core element. When people talk authentically and share images of possibility around such an important arena of inquiry and practice in widening networks of conversation, everyone benefits.

References

Bateson, G. (1972). *Steps to an ecology of mind.* New York: Bantam Books.
Bateson, G. (1980). *Mind and nature: A necessary unity.* New York: Bantam Books.
Brown, J., Isaacs, D., & World Café Community. (2005). *The World Café: Shaping our futures through conversations that matter.* San Francisco, CA: Berrett-Koehler.
Capra, F. (1996). *The web of life: A new scientific understanding of living systems.* New York: Doubleday.

Freire, P. (1998). *Pedagogia da autonomia: Saberes necessários à prática educativa* [Pedagogy of autonomy: Necessary knowings for an educational practice]. São Paulo, Brazil: Paz e Terra.
Freire, P. (2011). *Pedagogy of the oppressed*. New York: Continuum International Publishing Group.
Langer, E. (1997). *The power of mindful learning*. Reading, MA: Addison-Wesley.
Maturana, H. R., & Varela, F. (1987). *The tree of knowledge*. Boston, MA: Shambhala.
Maturana, H. R. (1989). *Ontologia del Conversar* [Ontology of conversing]. *Persona y Sociedad, III*(2): 9–28.
Maturana, H. R., & Verden Zöller, G. (1996). The biology of love. In G. Opp & F. Peterander (Eds.). *Focus Heilpadagogik* (n.p.). Munich: Ernst Reinhardt. Retrieved March 15, 2022 from https://bsahely.com/2018/01/16/biology-of-love-by-humberto-maturana-romesin-and-gerda-verden-zoller/
Maturana, H. R. (2002a). *A ontologia da realidade* [The ontology of reality]. Belo Horizonte, Brazil: Editora UFMG.
Maturana, H. R. (2002). *Emoções e linguagem na educação e na política* [Emotions and language in education and politics] (3rd Ed.). Belo Horizonte, Brazil: Editora UFMG.
Maturana, H. R., & Verden Zöller, G. (2008). *The origin of humanness in the biology of love*. Charlottesville, VA: Imprint Academic.
Maturana, H. R., Yánez, X. D., & Muñoz, R. M. (2015). *Cultural-biology: Systemic consequences of our evolutionary natural drift as molecular autopoietic systems*. Dordrecht: Springer Science+Business Media.
Mesquita da Silva, F., & Mesquita, M. (2016). *El World Café y la Generación de la Paz en Brasil* [The World Café and the Generation of Peace in Brazil]. In J. Brown, D. Isaacs, & Comunidad del World Café (Eds.), World Café: Construyendo Nuestro Futuro a través de Conversaciones Poderosas (Appendix). Lemoine Editores.
Mesquita da Silva, F. (2017). *Generation of peace dialogues: How the World Café approach to community understanding led to cultures of peace*. Fielding Graduate University dissertation. ProQuest Dissertations Publishing, 10601890.
Nicolescu, B. (2010). Methodology of transdisciplinarity—Levels of reality, logic of the included middle and complexity. *Transdisciplinary Journal of Engineering & Science, 1*(1), 19–38. Retrieved on August 2, 2013 from http://basarab-nicolescu.fr/Docs_Notice/TJESNo_1_12_2010.pdf
Schön, D. (1983). *The reflective practitioner: How professionals think in action*. New York: Basic Books.
Steier, F., Brown, J., & Mesquita da Silva, F. (2015). The world cafe in action research settings. In H. Bradbury (Ed.), *The Sage handbook of action research* (pp. 210–218, 3rd ed.). Thousand Oaks, CA: Sage.
Thompson, W., Steier, F., & Ostrenko, W. (2014). Designing communication process for the design of an idea zone at a science center. *Journal of Applied Communication Research*, pp. 1–19. London: Routledge. Retrieved March 15, 2022 from http://dx.doi.org/10.1080/00909882.2013.874570

Bunnell, P. (2020). *Palm Leaf*. Photograph.

The Love That Was Not Recommended:
Maturana's Biology of Love

Seiichi Imoto[1]

In Maturana's biology of love, love is the acceptance of the legitimacy of all existence, that is, of oneself, the others and the circumstances. Now, however, the legitimacy of all existence is in crisis. All the beings are crying: Don't deny me my existence! The rights to be (i.e., to exist or to survive) are ethical concerns, and ethics is based on the love that accepts their existential legitimacy. We need, hence, Maturanean love to protect the rights of all existence including us human beings. Maturana said, however, that he had no intention of recommending love. Why didn't he recommend the love? In this article, I would like to say: it seemed inevitable for Maturana not to recommend the love, judging both from the character of his biology, the biology of structure-determined systems, and from his own personal character. Geologically, we are living in the epoch of *Holocene*. We can say, however, we are actually living in the epoch of *Homocene,* in the epoch of human artificialness. In order to improve the conditions of the rights to be of all the beings, Maturanean love must be awakened and cultivated among all the people. We have to create a pedagogy of love on the basis of Maturana's biology of love.

Keywords: Maturana, love, legitimacy of existence, human rights, pedagogy of love

---

Introduction

In Maturana's biology of love, love is the acceptance of the legitimacy of all existence, that is, of oneself, the others and the circumstances (Maturana & Verden-Zöller, 2008, pp. 106, 120).

Now, however, the legitimacy of all existence is in crisis. António Guterres, Secretary-General of the United Nations, says: "Our world is in trouble. People are hurting and angry. They see insecurity rising, inequality growing, conflict spreading and climate changing. … We need to reform the world" (Guterres, 2017, pp. 24, 30).

All the beings are crying: Don't deny me my existence! The rights to be (i.e., to exist or to survive) including our human rights are ethical concerns. Ethics is based on the love that accepts their existential legitimacy (Maturana & Poerksen, 2004, p. 205). We need, hence, Maturanean love to save the world, to protect the rights of all existence including us, human beings.

Maturana said, however, that he had no intention of promoting (or recommending) love (Maturana & Poerksen, 2004, p. 207). Why didn't he recommend (or promote, urge, persuade) the love? This is my first question in this article.

Chile, Maturana's motherland, is a country that has one of the world's largest income gaps, and on the outskirts of Santiago, there are many poor people living in the slums, the poblaciónes. How did Maturana see those poor people? Did he accept their

---

1. Independent Researcher. Sapporo, Japan. Email: s-imoto@ca3.so-net.ne.jp

legitimacy as human beings and try to care for them? This is my second question in this article.

In this article, I would like to say that it seemed inevitable for Maturana not to recommend the love, judging both from the character of his biology, the biology of structure-determined systems, and from his own personal character.

Geologically, we are living in the epoch of *Holocene*. We can say, however, we are actually living in the epoch of *Homocene,* in the epoch of human artificialness, not of nature's naturalness. In order to reform or transform the world for the better and to realize our solidarity, Maturanean love must be awakened and cultivated among all the people. We have to create a pedagogy of love on the basis of his biology of love.

Maturana's Biology

It can be said that Maturana's biology is the biology of structure-determined systems, or of structural determinism. Agents external to them only trigger the systems but cannot determine them. A structure-determined system can also be called an operationally-closed system, and hence, a system of autonomy (self-governance) or a system of let-it-be (laissez-faire). Therefore, Maturana's biology can be called the biology of autonomy, or of let-it-be, of laissez-faire. The last sentence of the book *From Being to Doing* is just the expression of his biology of autonomy, which says: "All I am saying is: *We bring forth the world we live by living it.* Whatever we wish we should do" (Maturana & Poerksen, 2004, p. 208; italics in original).

Biology of cognition and biology of love are the two aspects (or derivatives) of the biology of structural determinism. Biology of love consists mainly in the domain of interactions (or relational domains) of structure-determined systems. Love spontaneously emerges when structural congruence has arisen among two or more structure-determined systems, and in this situation, they are in harmony or in resonance. If the systems are human individuals, they can reflect on themselves with the aid of language. Each individual can be liberated up to a being-for-itself (Für-Sich-Sein), which can enjoy a state in full awareness, responsibility and freedom. To act in full awareness of responsibility for the others is an expression of the love which accepts the legitimacy of their existence, and the basis of democracy in which an operationally-closed system is respected as an individual.

Maturana's Grounds for "Not to Recommend Love"

Maturana insisted:

> I would like to repeat once more: I have no intention whatever of promoting love but I do indeed insist that there can be no social phenomena without love. (Maturana & Poerksen, 2004, p. 207)

What is the reason why he does not promote (or recommend) love? The answer may be found in these statements:

> My view is that even so-called ethical laws and imperatives destroy the possibility of reflection: They remove the foundations of personally responsible action and demand submission; they are, at closer inspection, just another expression for tyranny. (Maturana & Poerksen, 2004, p. 49)

> Any attempt to persuade applies pressure and destroys the possibility of listening. *Pressure creates resentment.* Wanting to manipulate people stimulates resistance. (Maturana & Poerksen, 2004 p. 51; italics in original)

> Commandments of all kinds always tend to manoeuvre us fatefully close to the roles of missionaries and tyrants. (Maturana & Poerksen, 2004, p. 201)

Maturana did not want to be a tyrant nor a missionary nor did he want to create resentment and resistance in people. In this point, however, his reasoning is wrong. Even if he is a tyrant, his listeners are structure-determined systems. Whatever happens in them is determined by their structures, not by what he says to them. Maturana can only perturb them, hence, he cannot be a tyrant. He should have recommended the love to them without minding any consequence.

Maturana's Character

Maturana said:

> If I want to keep valuation and description clearly separate, all I have to do is to argue as clearly and as precisely as possible and state exactly what I mean and what I want to say. (Maturana & Poerksen, 2004, p. 207)

> Biology does not tell us what we must do, and as a biologist and therefore as a scientist I cannot tell anyone what to do. (Maturana & Poerksen, 2004, p. 206)

> Whoever I am talking to, I'm talking to as a biologist. (Maturana quoted in Von Foerster, 2002, p. 9)

He had been a biologist or a scientist; he just wanted to describe phenomena but did not want to evaluate them. He added further:

> I am not a revolutionary and I do not see myself as a man with a mission to change the world; I simply want to demonstrate how certain processes produce certain entities, that is all. (Maturana & Poerksen, 2004, p. 159)

> I myself was never a revolutionary. All I want is to do my job properly. (Maturana & Poerksen, 2004, p. 163)

He had not a disposition of a revolutionary who would want to transform the world politically. He just wanted himself to be a structure-determined system that acts simply in full awareness of its responsibility and freedom, of its autonomy and self-respect. He further said:

> The decisive thing is that observers must not be influenced by their own ambitions and the wish to achieve a specific result. This is precisely the reason why an observer can perceive anything at all;

> for whoever wants to see and understand something, must let it happen and let it show itself. The motto for the kind of perception that makes such understanding possible, and is founded on love, is: *Let it be!* (Maturana & Poerksen, 2004, p. 157; italics in original)

He just wanted himself to be an observer as described in this citation and wanted all other beings to act in the manner of "Let it be!" accepting their legitimacy of existence, namely, seeing them in love.

Even under Pinochet's dictatorship, he wanted to conserve his dignity, namely, his own autonomy and self-respect. He protected his own dignity as the following citation suggests:

> I decided to pretend in order to stay alive and to protect my family and children. At the same time, I tried to move and behave in such a way as to avoid endangering my dignity and my self-respect. I kept away from certain situations, respected the curfew, did not discuss certain topics in the university. When soldiers came and ordered me to raise my hands and to move up to the wall, I raised my hands and moved up to the wall. However, it was quite clear to me in those moments that the time would come when I would no longer be prepared to grant power to the dictator's regime. (Maturana & Poerksen, 2004, p. 173)

> I was and I am an uncompromising scientist who glided through the years of the dictatorship, always confident and careful to produce impeccable work without logical faults. That is all! (Maturana & Poerksen, 2004, p. 184)

Maturana's Attitude Towards the Poblaciónes

While he had been concentrating on his own job, how did Maturana see the people living in the poblaciónes, the people oppressed in poverty? He said:

> I am totally serious when I say: We always do what we want to do, even though we may claim to be acting against our will or to have been compelled to do something. In such cases we desire the consequences of our actions although we may not like what we are doing at the moment. (Maturana & Poerksen, 2004, p. 172)

Is this also the case with the people in the poblaciónes, with the oppressed, the excluded, the refugees, or with the people whose human rights are violated? He also said:

> Ethical concerns arise at the moment when self-awareness emerges and when, therefore, the possible consequences of one's actions for another human being of personal importance are consciously reflected. (Maturana & Poerksen, 2004, p. 205)

When he talked of the love as the acceptance of the legitimacy of all existence, did he take into consideration the people surviving in the poblaciónes as those of personal importance? I don't know the answer, but what he talked of his university students under the Pinochet regime is of help to guess:

> I was careful not to attack the government in any direct way or to campaign openly for some political end—that was not my thing. I never urged my students to go in a certain direction but I wanted to develop their capacity for reflection step by step. (Maturana & Poerksen, 2004, p. 176)

He did not act politically, which was not his thing. What he wanted each of his students to do was for each to develop his or her capacity for reflection, namely, to become a structure-determined system liberated in full awareness of responsibility.

I don't know whether he actually went down to the poblaciónes and talked together with the people there. I imagine, however, he would have shown, even in the presence of the people of the poblaciónes, the same attitude as was shown to his students.

The Courage to Act

I surmise Maturana is rather a man of words and description, not a man of valuation and political action. He was not a revolutionary to transform the world; he elucidated the world through the generative mechanisms that he abstracted. He lived in the world of let-it-be, laissez-faire, or nature's naturalness, and observed the relational processes of events, expressing them as generative mechanisms.

The time, however, has changed. The world has become the world of human artificialness: In structured-determined systems, their domains of composition and interactions are changing through various designs by humans. Holocene, the present geological epoch, can be even called Homocene now.

In this epoch, human rights are in crisis. Paulo Freire[2] wrote in his *Pedagogy of Hope*:

> What excellence is this, that manages to "coexist with more than a billion inhabitants of the developing world who live in poverty," not to say misery? Not to mention the all but indifference with which it coexists with "pockets of poverty" and misery in its own, developed body. What excellence is this, that sleeps in peace while numberless men and women make their home in the street, and says it is their own fault that they are on the street? What excellence is this, that struggles so little, if it struggles at all, with discrimination for reason of sex, class, or race, as if to reject someone different, humiliate her, offend him, hold her in contempt, exploit her, were the right of individuals, or classes, or races, or one sex, that holds a position of power over another? What excellence is this, that tepidly registers the millions of children who come into the world and do not remain, or not for long, or if they are more resistant, manage to stay a while, then take their leave of the world? (Freire, 2021, p. 98)

What could be more important than our human rights? We need love to protect human rights and to transform the world for the better. Maturana and Verden-Zöller described:

---

2. Paulo Freire (1921–1997) was born in Brazil and is called "the primary teacher of the poor" (Freire, 2016, p. 43). After the 1964 coup in Brazil, he moved in Chile and lived there for four years. His famous book *Pedagogy of the Oppressed* was written in this refuge in Santiago. Although I don't know whether Freire and Maturana had actually met each other somewhere in Santiago, if it had so happened, however, it must have been a time of a very fruitful dialogue for each other.

> Only if we want to conserve our *Homo sapiens-amans* condition and live accordingly, will our *sapiens-amans* lineage be conserved. (Maturana & Verden-Zöller, 2008, p. 133)
>
> We prefer to think that we human beings of the *amans* lineage have fallen asleep; let us now awaken. We human beings are not gods, but we are *Homo sapiens-amans*, and we are aware now that our fundament is the biology of love and intimacy. (Maturana &Verden-Zöller, 2008, p. 146)

We have to act now; we should have the courage to act to awaken us as *Homo sapiens-amans*, on the basis of Maturana's love. We need a pedagogy of love and solidarity grounded on the biology of love.

Freire wrote about Che Guevara's love in his *Pedagogy of the Oppressed*:

> I am more and more convinced that true revolutionaries must perceive the revolution, because of its creative and liberating nature, as an act of love. For me, the revolution ... is not irreconcilable with love. On the contrary: the revolution is made by people to achieve their humanization. What, indeed, is the deeper motive which moves individuals to become revolutionaries, but the dehumanization of people? The distortion imposed on the word "love" by the capitalist world cannot prevent the revolution from being essentially loving in character, nor can it prevent the revolutionaries from affirming their love of life. Guevara (while admitting the "risk of seeming ridiculous") was not afraid to affirm it: "Let me say, with the risk of appearing ridiculous, that the true revolutionary is guided by strong feelings of love. It is impossible to think of an authentic revolutionary without this quality. (Freire, 201, p. 89)

Che Guevara was born in 1928, the same year of birth as Maturana's. And now, Gabriel Boric[3] was elected the new president in Chile, surely with Che Guevara's love in his heart. I believe he has the courage to act on the basis of love. I earnestly hope he will transform Chile's human right status for the better.

Acknowledgments

I once met Humberto R. Maturana in Santiago, Chile. It was late September in the year of 2003. Locust trees were in full bloom. It was an unforgettable time of my life. Thank you so much for your friendship, Humberto! You were a great man.

References

BBC News. (2021, December 21). *Leftist Gabriel Boric to become Chile's youngest ever president*. Retrieved March 23, 2022 from https://www.bbc.com/news/world-latin-america-59715941
Freire, P. (2016). *Pedagogy of commitment*. New York: Routledge.
Freire, P. (2018). *Pedagogy of the oppressed* (50th anniversary edition). London: Bloomsbury Academic.
Freire, P. (2021). *Pedagogy of hope*. London: Bloomsbury Academic.
Guterres, A. (2019). Address to the General Assembly, UN [January 3, 2017]. In United Nations Association of Japan (Ed.), *New Today's guide to the United Nations* (pp. 23-40). Tokyo: Sanshusha. Retrieved March 23, 2022 from https://www.un.org/sg/en/content/sg/speeches/2017-01-03/secretary-generas-address-staff

---

3. Gabriel Boric, a 35 years-old leftist, won the presidency of Chile at the end of the last year. His regime will begin from March 2022. Mr. Boric vowed to tackle problems against human rights such as economical / political inequalities and corruptions. (BBC News, 2021)

Maturana, H. R., & Poerksen, B. (2004). *From being to doing: The origins of the biology of cognition.* Heidelberg, Germany: Carl-Auer Verlag.
Maturana, H. R., & Verden-Zöller, G. (2008). *The origin of humanness in the biology of love.* Exeter, UK: Imprint Academic.
Von Foerster, H. (2004). Preface. In H. R. Maturana & B. Poerksen, *From being to doing: The origins of the biology of cognition* (p. 9). Heidelberg, Germany: Carl-Auer Verlag. (Preface was written in 2002)

Bunnell, P. (2021). *Distinctions and Domains.* Photograph.

Bunnell, P. (2016). *Natural Drift*. Photograph.

Maturana's Path of Objectivity-in-Parenthesis

Raul Espejo[1]

This article discusses the explanatory paths that Maturana calls objectivity-without-parenthesis, or the path of transcendental objectivity, and the path of objectivity-in-parenthesis, or the path of constituted objectivity (Maturana, 1988). I relate these views to Black Box descriptions and operational descriptions of organizational systems (Espejo & Reyes, 2011). The most significant implication of this distinction is that while Black Box descriptions are focused on the relational complexity of the social system with its environment, the operational descriptions are focused on the complexity of the relationships producing these organizational systems from the multiple stakeholders' viewpoints, accounting for aspects such as respect, trust, collaboration, cooperation and in more general terms to the emotions of love constituting these relationships. It is argued that this second-order cybernetics perspective is complementary to the first order, Black Box perspective, and adds to our understanding of Ashby's requisite variety (Ashby, 1964) and Beer's viable system model (Beer, 1979).
Keywords: Maturana; objectivity-without-parenthesis; objectivity-in-parenthesis; organizational systems; viable system model; requisite variety

Introduction

In this article I explore from the perspective of organizational cybernetics Maturana's two explanatory paths, namely the transcendental ontologies of *Objectivity*, that is *objectivity-without-parentheses*, and *(Objectivity)*, that is, *objectivity-in-parenthesis* (Maturana, 1988). In my work I relate these paths to a Black Box description and to an Operational description of organizational systems respectively (Espejo & Reyes, 2011). The most significant implication of this distinction is that the Black Box, or Objectivity description is focused on the complexity of outputs to the environment starting from the organization's transformation of its inputs, while the operational description, or (Objectivity) is focused on the complexity of the relationships between its multiple internal and external stakeholders producing the outcomes of the organizational system from its relationships while accounting for aspects such as respect, trust, collaboration, and cooperation. In more general terms this entails accounting for the emotion of love, or its lack, in the constitution of these relationships. This is a second-order cybernetic perspective that is complementary to the first order Black Box perspective, and it helps to developed a significantly more comprehensive understanding of conversations in social systems.

 My discussions are focused on people and organizational systems and are influenced in general by Maturana's work and his view that observers are living

---

1. Email: r.espejo@syncho.org

systems as well as emotional and conversational beings whose cognitive abilities are altered by their biology and languaging, Further, organizational systems are constituted by observers; "everything said is said by an observer to another observer that could be him- or herself" (Maturana, 1988, p. 28).

My arguments focus on two of his articles: "Reality: The Search for Objectivity or the Quest for a Compelling Argument" (Maturana, 1988) and "Autopoiesis, Structural Coupling and Cognition: A History of These and Other Notions in the Biology of cognition" (Maturana, 2002). The first article gives, from my perspective, an insight about his ontological and epistemological views of people contributing to the constitution of social systems and the second helps to understand the concepts of organization, structure, relationships and identity in social systems rather than in biological systems. However, as Maturana has said, organizations are social systems and not living systems, though many people have tried to apply Maturana's original concepts about living systems to organizations.

In social systems the complexity of the operational domain depends largely on the complexity of autonomy, cognition, and communications, all of which are aspects of the *structural couplings* among the participants. The complexity of this domain is orders of magnitude larger than that of people's distinctions in their *informational domain,* that is, in the domain of referring to these processes.[2] For effective participation in change processes we need to take into account the complexity of structural couplings (Maturana, 2002). I emphasize that because people operate in networks, the operational domain of their actions needs to account for the complexity of their interactions.

Maturana states that notions such as complexity and chaos are evocative metaphors for the reflections of an observer and argues that these notions do not reveal the processes involved in the constitution of a social system. However, I argue that being aware of sources of complexity enables this system to configure its interactions. For a social system, *structural recursion* is replication of its structure within the contained autonomous units (Beer, 1979). This recursion makes it possible for the system to cope with proliferating environmental complexity. In other words, the components of a social system with structural recursion are autonomous sub-systems each structurally determined and operationally closed, as described in Figure 1.

Not recognizing structural recursion implies failure to see the architecture of social systems and, from the perspective of an observer, implies sticking to a transcendental ontology, which fails to recognize the components' autonomy. In a phenomenological sense, a complex structure determined system is generated as the system's components respond with autonomy to the environment's larger number of possible states. I relate this complexity to Ashby's concept of variety: the number of possible states of a situation, which he offered as a measure of complexity (Ashby, 1964).

---

2. The distinction between operational and informational domains recurs throughout the arguments of this paper (Espejo, 2020).

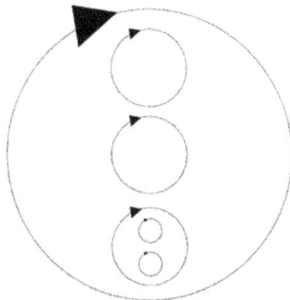

Figure 1. Structural Recursion (adapted from Beer, 1979, p. 315). Self-organizing autonomous organizations are indicated by the circular arrows. Smaller autonomous organizations are shown as viable entities within the larger autonomous organization.

From the perspective of an observer, autonomy relates to (Objectivity) rather than Objectivity. People who conceive their world according to Objectivity get no benefit from observing the structural recursion of autonomous components in the organizational system. Thus, my view is that complexity is more than an evocative metaphor; it is a necessity for structure determined systems in demanding environments. Variety is central to structural determination and to the structural couplings of organizational systems, and requisite variety is the amount necessary for a viable response in a complex environment. The organization's relationships of legitimation, domination, and signification (Espejo et al., 1996, p. 70) constitute their bodyhood and make requisite variety a central concept in the formation of its operational domain, as is discussed later in the paper.

Emotional upheavals that may lead to the mutual destruction of participants in a cognitive disagreement is an inevitable consequence of their operation in the explanatory path of Objectivity; that is, in the path of not seeing participants as autonomous beings in need of mutual respect. Disagreements in this explanatory path constitutively entail mutual negation and produce existential threats. The only way to escape such an emotional trap is to move to the explanatory path of (Objectivity), which includes autonomy, conversations and correction of relational variety imbalances. Recognition of the other cannot take place through reason, it can only take place through stable conversations; that is, through languaging and emotioning in the mutual acceptance of each other.

Organizations Arise Through Languaging Relationships

Language is a manner of living together in a flow of coordination of coordinations of consensual behaviors, or doings, that arise in a history of living in the collaboration of doing things together (Maturana, 2002, p. 24). People may language effectively in their *structural coupling* if they possess the necessary complexity or variety. The continuous interlacing of coordination of coordinations of doings with emotions is what Maturana calls conversations. Organizational systems emerge from networks of these conversations.

Figure 2 shows the coordination of actions of two people sharing the same music as a common source (Espejo, et al., 1996). This is a dance in the operational domain of their interactions. They learn in these interactions, and possibly they coordinate this coordination of their actions if they exchange information about what they are doing and how to do it. If they language this "how to do it," and if their exchanges are braided by the mutual emotion of acceptance, they are in a conversation. On the other hand, if they are good dancers and share the music, they don't need to coordinate their coordination of actions: It just happens through consensual coordination. The structural coupling happens without information exchanges and the dancers express their knowledge in their adequate performance. This kind of communication—without explicit recourse to language between the actors sharing a common context—is most powerful in managing complexity in organizational systems; Conant (1979) calls it *communications without a channel*.

Figure 2 Two dancers coordinating by sharing music as a common changing environment triggering their consensual coordinations. This is referred to as "communications without a channel" (Espejo et al., 1996, p. 73)

If what takes place along a particular course of recurrent interactions between two or more living systems is the expansion of an initial domain of consensual co-ordination of actions, without language between them, in Maturana's terms those living systems have established what he calls a domain of "consensual co-ordination of actions" (Maturana, 1988, p. 33). Following Conant (1979), I relate this domain of consensual coordination to the coordination of actions (without channel capacity) in the recurrent interactions of the living systems. All that is needed for these systems is that at their first encounter they have the necessary structural configuration (i.e., resources and relations) for recurrent interactions between them to take place. However only if structural plasticity in the domain of their interactions is present, and the initial structure allows them to conserve organization and adaptation can their initial consensual coordinations become a domain of recurrent interactions.

Towards a Path of Objectivity-in-Parenthesis

As already explained, related to living in language, Maturana refers to two ontological domains (Maturana, 1988); the domain of constitutive ontologies (Objectivity) and the domain of transcendental ontologies Objectivity (Figure 3).

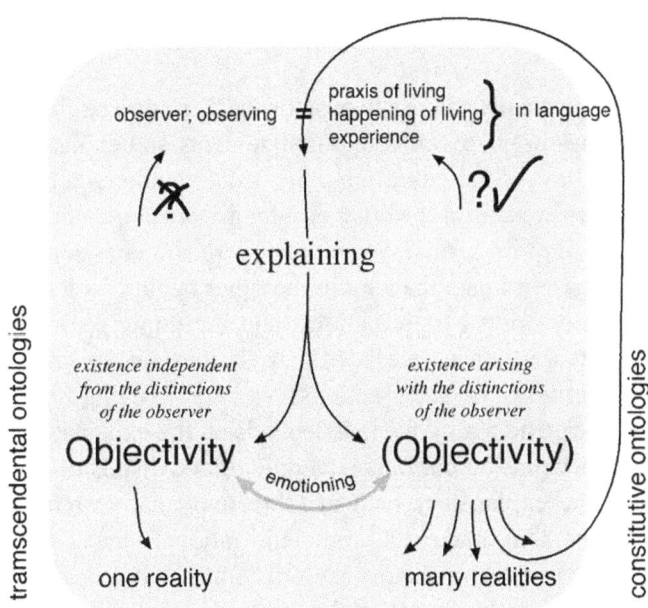

Figure 3. Ontological Domains as Drawn by Maturana in Many Workshops
(As redrawn by P. Bunnell in *CHK*, 2004. Original published in Maturana, 1988, p. 32)

An observer who follows the explanatory path of (Objectivity) realizes that he or she lives in a *multiversa* of "many different, equally legitimate, but not equally desirable, explanatory realities, and that in it an explanatory disagreement is an invitation to a responsible reflection of coexistence, and not an irresponsible negation of the other" (Maturana, 1988, pp. 31–32). In understanding this the observer also realizes that other observers are also autonomous beings who live in different, yet legitimate realities.

An observer in the domain of transcendental ontologies claims explanations are validated by their reference to entities which exist independently of its actions; matter, energy, gravity and so on. In this ontology people disagree about reality, thinking their reality is the right one and other people are wrong.

Whether any particular observer operates in either the transcendental or the constitutive ontologies depend on whether or not he/she accepts different domains of actions and cognition as valid. The acceptance, in turn, is grounded in an emotional choice of either trust or mistrust and the attendant desire for certitude. Whenever the observer operates assuming Objectivity, there is an implicit operational acceptance of any observation as a representation of Reality. It is only when the observer accepts the question about how we explain biological or social phenomena (Figure 3) that the possibility of two explanatory paths, namely of (Objectivity) and Objectivity appear, and it is only then that it is possible for the observer to reflect upon their epistemological and ontological implications. Whether the observer follows one explanatory path or the other, however, does not depend on a rational argument—it depends on preferences and inner disposition to implicitly or explicitly accept and take one or the other of these two possible starting conditions; Objectivity or (Objectivity).

In daily life, we normally move unconsciously from one explanatory path to the other in the manner we argue to validate our statements and explanations, and we do this according to the flow of our emotioning in our interpersonal relations and desires. Thus, if in a discussion we accept the other as a legitimate and autonomous individual, and we don't impose our views, then we are treating the other as someone operating in a different but equally legitimate reality. On the other hand, if we assert our position, or force the other to perform certain actions, and are doing so with the implicit or explicit justification that we have access to the truth, then we are acting according to a transcendental Objectivity. In such actions, we are denying the autonomy and legitimacy of the other. The reality we live depends on the explanatory path we adopt, and that this in turn depends on our understanding of and disposition to (Objectivity).

As a result, in the explanatory path of Objectivity the search for reality is the search for conditions that make an argument rational, and, hence, undeniable. However, an observer in the explanatory path of (Objectivity) is aware that, although emotions do not determine the operational coherences of any domain of reality, they determine the domain of operational coherences of rational arguments. It is apparent that when an observer, in a position of authority, shouts to someone he or she is likely to be operating in a path of Objectivity.

In the discussion of organizational systems, because of structural determination, I assume observers are aware of Objectivity and (Objectivity). However, it is apparent that people in organizations often follow the path of Objectivity, The two paths trigger different organizational forms. I am proposing a further possibility, that is to operate in the confluence of these two paths as this enables the humane operation of complex organizations. Unfortunately, a humane operation is not the most common. In daily life we usually operate in the explanatory path of Objectivity, which in most practical situations of moving from one place to another or doing routine tasks, is not only acceptable but expedient. However as a consequence we easily become blind to the importance of our emotioning in shifting us between the two paths, and as a consequence easily create, and then become trapped in hierarchical structures, where rationality supersedes respect, acceptance and appreciation for the others. It is only as we become aware of the biology and emotions of the observer and operate in the explanatory path of (Objectivity), that we become aware that every rational system in which we operate is grounded in basic premises that we adopt according to our emotioning. Respect, acceptance and more generally as proposed by Maturana, love, is the emotion that constitutes social phenomena. Respect, acceptance, and love are emotions that specify the domain of actions in which living systems co-ordinate their actions and develop their autonomy (Maturana, 1988).

Organizational Systems Are Created by Human Beings in Conversation Networks

Any network of conversations in which people operate in the mutual acceptance of each other may constitute, through their relationships, a social system. Thus a family, a community, a political party, a sports group or a cohesive group of friends may all be systems of co-ordinations of actions in language, and as such comprise networks of conversations realizing social systems. The richness and resilience of such networks depends on how the people involved operate with requisite variety, and do so in mutual acceptance. As a result, and regardless of our awareness of this situation, we move in daily life through networks of conversations, entering and leaving social systems according to whether, in the flow of languaging and emotioning, our behavior entails accepting or rejecting coexistence in mutual acceptance.

As claimed in the introduction, the explanatory path of Objectivity corresponds to a Black Box (BB) description of an organization whereas the path of (Objectivity) corresponds to an operational description. The BB and operational domains of coordination of actions affect our emotional dispositions and the way we construct our interactions with others and thereby create domains of mutual acceptance. In the BB domain we don't recognize shared domains, we just operate in different domains. On the other hand, in the operational domain an observer may claim that social phenomena are taking place when two or more people, in recurrent interactions, follow an operational course of mutual acceptance thus forming a social network. It entails mutual acceptance and constitutes social phenomena (Maturana, 1988).

Maturana's arguments about Objectivity and (Objectivity) are particularly useful for explaining aspects of autonomy and complexity in organizational systems. They are not either-or arguments, together they are relevant to the constitution of organizational systems.

People who understand their contribution to organizations in a mood of Objectivity assume that their understanding of the situation defines a transcendental reality that they discern while the views and emotions of others might be considered to be misleading or wrong. This stance is a characteristic of people operating in a hierarchical organization where the operations are made efficient through restricting the autonomy of workers. In this stance bosses in effect are defining the reality of the organization and the tendency will be to take the purposes of the organization as those visualized and defined by these bosses. The role of conversations and inclusion of others' perspectives and even insights in the definition of these purposes is likely to be minimized and the organization is likely to be seen as a trivial machine of inputs and outputs (Espejo, 1994). In other words, Objectivity will overemphasize unilateral power. People with this perspective, even if they believe otherwise, are describing organizations as Black Boxes. While this is unlikely to be a completely definitory description, it provides a common description of the sort we often encounter in daily life.

On the other hand, if people understand organizations in a stance dominated by (Objectivity) they will recognize the organization as constituted by conversations of mutual respect, in an emotion of love, which I have called an *operational description of organizations*. When the constitution of an organization is dominated by (Objectivity), people do not impose their purposes but construct them through conversations. I illustrate this form of organizational constitution later in this article.

It is in this sense that I'm making a connection between Maturana's ontological domains (Figure 3) and Beer's viable system model (VSM; Figure 5, below). Most significantly, central to my argument is that Beer's views about autonomy and complexity are strengthened through Maturana's ontological views.

Input-output (BB) descriptions (Figure 4) of social systems emphasize the power of information in organizational adjustments rather than the power of communication. Communication entails structural coupling among structurally determined systems which thus results in change and adaptation. Input-output (BB) descriptions can be related to first-order cybernetics. Operational descriptions, on the other hand, can be related to second-order cybernetics and to an appreciation of autonomy and the complexity of embodied relationships. Both BB and operational descriptions are necessary and complementary. The first puts the emphasis in the environmental constraints that orient how an organization handles complexity. The second puts the emphasis in people's conversations, language and emotions, that constitute the purposes of the organization. The first is about inputs and outputs, the second is about relationships. I see in this distinction the epistemological and ontological contributions of Maturana to social understanding of organizations as embodied processes. In what follows I will illustrate this complementarity with reference to Beer's VSM.

The BB description is focused on input-output requirements for managing the complexity of an organizational system. Espejo (2020) gives details of the structural requirements for managing the complexity of a situation with requisite variety. However, in organizational systems, these requirements should be the outcome of a constituted ontology of conversations of (Objectivity), This is the operational description of a system focused on relationships, in which people contribute to its constitution through language, emotioning and, following Maturana, through their braiding in conversations.

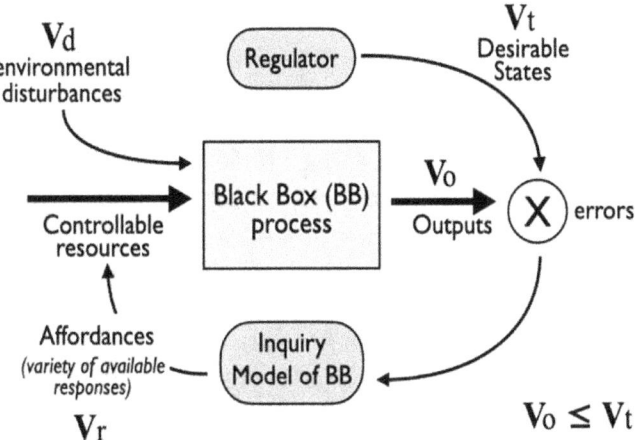

Figure 4. Management of complexity and requisite variety as represented by a Black Box description of an organizational system (after Espejo, 2022). V stands for variety and $V_o \leq V_t$, is a condition for requisite variety, that is, the variety of outputs, should be less/equal than the variety of desirable states.

In the operational description the complexity of the relationships shown in Figure 5, and discussed below, are the outcome of (Objectivity), and arise from multiple perspectives accordingly. These relationships define the bodyhood of the organizational system.

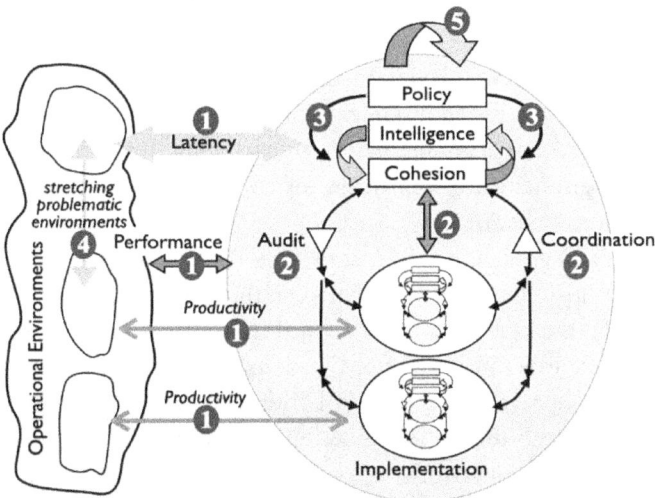

Figure 5. An adaptation of Beer's VSM with a focus on relationships; see text for an explication of the numbered relationship for creating, regulating and producing an organizational system's purposes (after Espejo 2008, 2022).

1. *Relationships of Performance*—Relationships of performance are knowledge embodied outcomes observed by observers who distinguish the organizational system behaving more or less adequately, with coherence, in its structural couplings with the environment. To a large degree performance is bodyhood driven; it is based on a history of structural couplings. Productivity is the outcome of actors' latency relationships of the system with stakeholders offering future possibilities and productivity relationships is the outcome of conversations between system's actors and external stakeholders focused on today's possibilities.
2. *Relationships of Cohesion*—Relationships of cohesion are those conserving cohesion between all the actors constituting the system (e.g. policy makers, experts, administrators, employees in general). In the extreme the governance of these relationships may follow conversations imposing hierarchical relationships of Objectivity. Alternatively, they may encourage relationships grounded in (Objectivity), thus supporting autonomy at each level of the organization. These are conversational processes for building trust and coordination of actions, far beyond the traditional executive information systems.
3. *Relationships of Policy Making*—Relationships of policy making take place between policy makers, inside actors and outside agents. Inside actors are focused on current operations inside the system (cohesion actors) and outside agents are focused onto the future (intelligence actors). The policy-makers' role is orchestrating conversations between both viewpoints in order to create desirable system purposes and values. These actors should recognize which conversations are necessary for the creation of values and purposes. More than defining policies, a key role of policy makers is working out who are the people that should participate in conversation policy debates. This is illustrated further in the section below.
4. *Relationships of Inclusion*—Relationships of inclusion take place between policy makers and the owners of the organizational system, that is, the people to whom these policy makers are accountable for. In particular, these relationships of inclusion and legitimacy are significant for constructing the social ownership of the organizational system.
5. *Relationships of Organizational Citizenship*—Relationships of organizational citizenship are those that take place between the system actors and those providing normative context to their activities. These are people who act as guardians of societal values in society, such as families, community, political parties, sports groups or even a group of friends. These relationships provide a context that enables co-ordination of actions in language. Through their emotional intertwining they create networks of conversations to the extent that the people involved operate in mutual acceptance. We move in daily life through networks of conversations, entering and leaving social systems according to whether the flow of languaging and emotioning entails accepting or rejecting coexistence in mutual

acceptance. Bringing this dynamic specifically into the operation of an organizational system arises through relationships of organizational citizenship.

Production of Organizational Purposes Through Stakeholder Relationships[3]

As a conclusion, I want to discuss further relationships of policy making (Relationship 3 in Figure 5) to illustrate the production of purposes and the implication of this production in the boundaries of organizations.

Rather than a single viewpoint defining the purposes of the social system, in Maturana's paradigm clarification of purposes follows the idea of (Objectivity). Clarification of a system's purposes requires more than just statements about its missions and goals. Rather it requires correcting variety imbalances between the views of external (environmental) and internal (organizational system) stakeholders, creating stable meanings through well-resourced and steered conversations between Policy, Intelligence and Cohesion as depicted in Figure 5.[4] These are systemic functions necessary for an effective policy process. Central to all this is that policy is the outcome of conversation between people rather than a matter of unilateral statements.

Systems' purposes evolve from conversations. Leadership needs to manage the interactions between those contributing to these purposes. Among other aspects these conversations contribute reflexively to the clarification of their roles. Some stakeholders will contribute to producing the system's products and others to dealing with environmental stretching. Purposes emerge from steering value driven conversations such as the pursuit of truth, sharing of knowledge, freedom of thought, rigorous, reasoned arguments, listening to alternative views, the impact of own views on others, and commitment to ethics, such as rule of law, democracy, human dignity, human rights, good governance and sustainable development. These values, as adhered to by organizational leaders, are proposed as drivers for stakeholders' interactions. In the context of value driven conversation, stakeholders are in better disposition to contribute to balanced relationships towards creating systemic policies. Though conversations concerning values may be initially loose, they should contribute reflexively to the in-depth clarification of purposes. Indeed, in social systems, these conversations depend largely on consensual coordination based on structural determination and structural coupling and to a lesser degree in information.[5] Those contributing to these processes will support through their reflexive interactions the production of the system's purposes (e.g. identity). An important role of policy makers is to balance these contributions.

Processes for the creation of policies should ground the system's development in the mechanism underpinning these conversations. Whatever the outcomes, these are grounded in correcting variety imbalances in conversations. Following Maturana, the

---

3. An application of this idea of purpose production to higher education is in Espejo and Holtham (in press).
4. This clarification requires variety balanced conversations between intelligence and cohesion viewpoints.
5. This is the distinction between operations and information used in different parts of this article

quality of these conversations, should be driven by a constitutive ontology of (Objectivity).

Though structural coupling may sound mechanical, it is a basic process that has the potential to enable people in conversation to coordinate their relationships and avoid unchecked controls and abuses of power. Policy interactions in organizations need to be driven by caring networks. These networks are necessary to embody a robust organization. A structure that prevents respectful and ethical interactions destroys what people think and do and is thus inadequate. Adequate structure, on the other hand, consist of recognizing that what each individual deserves is beyond unilateral statements of wishful thinking and should be the outcome of balanced conversations in the policy system (Espejo, 2022). Such a structure is encouraged by ethical considerations starting with policy makers, and from there expanding to integrating stakeholders in a manner that leads towards a collective of people sensitive to the system's long term. In the end, fairness and quality emerge from properly regulated interactions, reflecting the values and purposes of the related people in a wider environmental context.

An implication of the above discussion is that the boundaries of an organizational system are the outcome of conversations and therefore emotional in nature. Naturally, stakeholders constitute varied networks of co-ordination of actions. This is reflected in the dashed circles of figure 6 below; indicating that the boundaries of an organizational system are fluid. This issue of *liquid boundaries* (Bauman, 2000) is developed in two of my papers (Espejo, 2008, 2020). In particular the 2020 paper offers a methodological extension of this fluidity, through enterprise complexity networks.

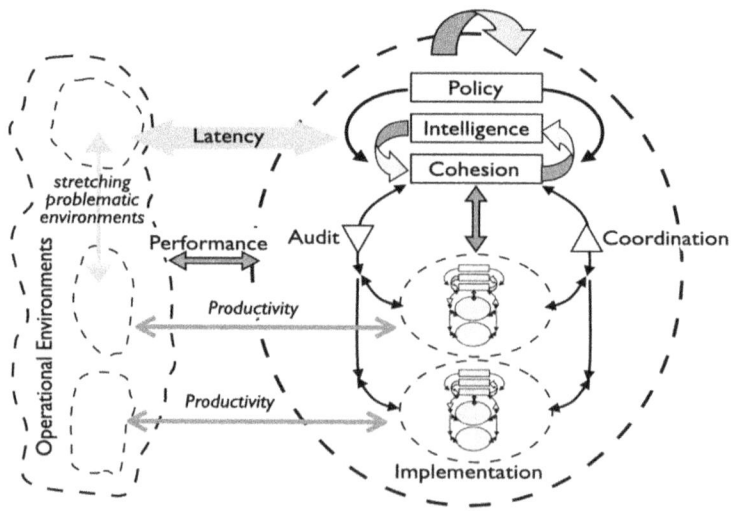

Figure 6. A network of enterprises embodied as a social system with liquid boundaries (after Espejo, 2008, 2020)

In summary, Maturana provides the foundations of his constitutive ontology in his paper of 1988. He argues that "a social system is a closed system that includes as members all those organisms that operate under the emotion of mutual acceptance in the realization of the network of co-ordination of actions that realizes it" (p. 70). My arguments about balanced interactions make explicit that they should be the outcome of (Objectivity), wherein multiversa are constituted based on language and emotions. And, unless participants overcome fragmentation, networks will fail to produce integrated outcomes, purposes and values, in which case they will have to accept rational arguments of objectivity and power. The issues that are considered to be of concern will be bounded by the limits determined by the transcendental reality of those in power. Thus, as Maturana would argue, boundaries of a social system appear in the behavior of its members as they include or exclude others from a particular network of co-ordination of actions. People usually justify this with rational arguments from the perspective of the explanatory path of Objectivity as a social boundary becomes explicit in language. Consequently conversations about purposes are far more effective than rational arguments as they enable correcting variety imbalances between actors for effective policy processes, and this requires managing emotions. The presence of emotions in our conversations make explicit that the boundaries of a particular social system require us to manage more than rational arguments.

This is where the rational underpinning of science, grounded in Objectivity becomes a difficult path to follow in policy processes. If the management fails to take into account adequate variety, then stakeholders, particularly leading ones will have to move out of the path of Objectivity into the path of (Objectivity) and take responsibility for their actions. If we are in the path of coexistence in (Objectivity), the situation is different because stakeholders will be increasingly aware of the many different domains of reality in which they operate, as well as of the emotional grounding of their ethical concerns. This is a situation where stakeholders will benefit by structuring their relevant situations from different cognitive domains; blending rational and emotional domains. Managing their interactions may benefit of the conversational tools as proposed by Gordon Pask, and others (Pask, 1979; Winograd & Flores, 1986; Scott, 2021). The many domains of reality or cognitive domains that we bring forth as we explain our praxis of living in the explanatory path of (Objectivity) appear as necessary to incorporate the benefits of Beer's, Maturana's, Pask's and also Flores's works.

Let's bring forth organizational systems that support living and working together from the path of objectivity in parenthesis, the emotion of love, the power of conversations and the correcting of variety imbalances in their development and productivity

Acknowledgements

Pille Bunnell's clarification of this manuscript cannot be underestimated. I thank her invaluable contributions through our conversations in the spirit proposed by this paper, over an extended period of time. I'm also grateful to Manuel Manga, whose reading of the original manuscript, to reduce its size, was additionally clarifying and illuminating.

References

Ashby, R. (1964). An introduction to cybernetics. London: Methuen & Co, Ltd.
Bauman, Z. (2000). *Liquid modernity*. Polity Press, UK.
Beer, S. (1979). *The heart of enterprise*. Chichester, UK: Wiley.
Bunnell, P. (2004). Reflections on the "Ontology of Observing." *Cybernetics and Human Knowing, 11*(4), 72–84.
Conant, R. (1979). Communications without a channel. International Journal of General Systems, 5, 93–98.
Espejo, R. (1993). Domains of interaction between organisation and environment. *Systems Practice, 6*(5), 517–525.
Espejo, R. (1994). What is Systems Thinking. *System Dynamics Review, 10*(Summer-Fall), 199–212.
Espejo, R. (2008). Observing organisations: The use of identity and structural archetypes. *International Journal of Applied Systemic Studies, 2*(1/2), 6–24.
Espejo, R. (2020). The enterprise complexity model: An extension of the viable system model for emerging organisational forms. *Systems Research and Behavioral Science., 1*, 1–17.
Espejo, R. (2022). Cybersyn, big data, variety engineering and governance. *AI & SOCIETY.* https://doi.org/10.1007/s00146-021-01348-0
Espejo, R. & Holtham, C. (in press). Practical wisdom for addressing contested problems. In I. Perko, R. Espejo, I. V. Lepskiy, & D. Novikov (Eds.), *Systems approach and cybernetics, engaging the future of mankind*. Proceedings of World Organisation of Systems and Cybernetics, 2021. Springer Nature.
Espejo, R., Schuhmann, W., Schwaninger, M., & Bilello, U. (1996). *Organizational transformation and learning*. Chichester, UK: Wiley
Espejo, R. & Reyes, A. (2011). *Organizational systems: Managing complexity with the viable system model.* Springer.
Maturana, H. R. (1988). Reality: The search for objectivity or the quest for a compelling argument. *Irish Journal of Psychology, 9*(1), 25–82. (Issue on constructivism)
Maturana. H. R. (2002). Autopoiesis, structural coupling and cognition: A history of these and other notions in the biology of cognition. *Cybernetics & Human Knowing, 9*(3-4), 5–34.
Maturana, H. & Varela, F. (1992). The tree of knowledge (rev. ed.). Shambhala.
Scott, B. (2021). *Cybernetics for the social sciences*. Leiden, Netherlands: Brill Publishers.
Pask, G. (1970). The meaning of cybernetics in the behavioural sciences (The cybernetics of behaviour and cognition; extending the meaning of "goal"). In J. Rose (Ed.), *Progress in cybernetics* (pp. 15–44). Gordon and Breach Science Publishers.
Winograd, T., & Flores, F. (1986). *Understanding computers and cognition: A new foundation for design*. Norwood, NJ: Ablex.

Bunnell, P. (2010). *Ripples*. Photograph.

Humberto Maturana: Using His Biological System in the Social Domain

Hugh Gash[1]

I was familiar with constructivist thinking in Dewey, Piaget, and von Glasersfeld when I discovered Humberto Maturana's work. I used Maturana's idea of objectivities in parenthesis in intervention studies in elementary schools focused on gender stereotypes and children with learning difficulties or from other countries. Recently I wrote about thinking about sustainability and climate change and hoped that *objectivities* captured in the phrase *consensual communities* might facilitate dialogue between groups with different views. Yet throughout there is always the tension between alternative objectivities. I am concerned about how to sustain mutual respect in conversations between individuals and consensual communities that have strongly opposing views.
Keywords: Maturana, objectivity in parenthesis, stereotypes, education, populism, climate change

Introduction

This paper is an account of my journey to and with constructivist thinking. In it I first met Jean Piaget and John Dewey and met the powerful idea that knowledge might be better considered a process of knowing. Appreciating the implications for this emerged as Ernst von Glasersfeld and Humberto Maturana's ideas provided ways of thinking about thinking. In what follows I discuss these ideas and ways I used them in intervention studies with children and more recently in thinking about the problems and tensions that arise between individuals and groups with opposing views. In the sections that follow I describe Maturana's influence on my work and then in the later sections of the paper I discuss limits to tolerance and deep disagreement. Here the tensions are well known and perhaps the stated willingness to reconsider, to question one's own, and my own, thinking provides a way to allow the journey to continue so as to avoid entrenched positioning.

Behaviorism in the early 20th Century had taken the view that the scientific study of psychology required observable behavior. In the 1950s this changed as though the limitations of restricting science to what can be seen and described were being realized across social science domains. For example, two publications about language in 1957 seem like a watershed for cognitive psychology signaling the end of the hegemony of learning theories in psychology. B. F. Skinner published *Verbal Behavior* (1957) and Noam Chomsky published *Syntactic Structures* (1957) introducing an account of language and it's acquisition that emphasised the importance of (unobservable) models of grammar. Chomsky's linguistics and Piaget's (1954, 1970) psychology offered viable alternatives to behaviorism, highlighting the

---

1. School of Human Development, Institute of Education, Dublin City University, Ireland
   Email: hugh.gash@spd.dcu.ie

importance of modelling individuals' cognition. In the domain of economic theory Ivan Moscati (2016) has recently described research into how the psychology of human behavior and human agency played a significant role in economic behavior in the middle of the 20th century.

Humberto Maturana and Jean Piaget approached cognition from a biological context. I attended Maturana's workshop presentations in Dublin in May 1986 and December 1990 and found they helped me to understand his signature iterative style. In my formative years the theories of John Dewey, Jean Piaget, and Humberto Maturana each required learning and becoming familiar with their individual approaches and terminologies to understand the substance: Each theory also significantly influenced my writing and my work. The opportunity to read John Dewey in some depth as a graduate student studying educational psychology led to my noticing similarities in the accounts of cognitive processes and development in Dewey's and Piaget's theories. Each emphasised content and process in different terminology. Piaget emphasised assimilation and accommodation as processes involved in constructing ever more complex cognitive structures. Dewey had a similar account of the experimental nature of thinking and emphasised how: "The nature of the problem fixes the end of thought, and the end controls the process of thinking" (Dewey, 1971, p. 15). This was written up later as a postdoc working with Charlie Smock's Piaget-based Follow-Through Program at the University of Georgia (Gash, 1974). It was there I met Ernst von Glasersfeld who also played a significant role in my professional development and my ideas about constructivism, ideas that soon afterwards led to Maturana. A key idea for me in both von Glasersfeld's work and Maturana's was the epistemological implication of realism that up to then had escaped my attention.

Ernst von Glasersfeld (1974) first used the phrase *radical constructivism* (RC) to outline the epistemological consequences of taking constructivism seriously. In Buffalo, reading *Quest for Certainty* (Dewey, 1929) I had experienced a moment of insight into this epistemological issue when Dewey discussed scientific knowledge as "the consequences of directed operations form the objects that have the property of being known" (pp. 86–87). Glasersfeld's (1974) paper was designed to highlight the implications of appreciating that "reality" and "truth" cannot be guaranteed by a process involving comparison between inside and outside. What is experienced as outside must be based on a comparison between ongoing experience and prior understandings: We cannot go beyond our sensory interface. On this basis Glasersfeld (1974) argued that what is constructed should be judged, not by truth, but by being viable in terms of the stability of the constructions used to describe regularities in experience. This stability is constrained by intra-individual consistency in that we have to make sense to ourselves, and by inter-individual consistency requiring us to make sense in our social groups.

The Follow-Through Program at the University of Georgia was Piagetian so it was natural to think about cognitive construction in an educational context and in particular, a mathematical one. Piaget's approach to thinking was grounded on children's emerging understandings of classification, seriation and the conservation of various concepts associated with mathematics such as number, length, area, and volume. All these concepts concern objects or phenomena that are clear and certain. If we have two green squares and three green circles, we have more circles than squares, but more green things than circles. There is usually no doubt about the identity of squares and circles, nor about there being more circles than squares. The confusion arises asking young children (say 5 years of age) a question that requires switching the categories (mentally) and comparing the circles with the superordinate category green. There are more circles than squares but there are more green things than circles.

Well known objects and topics, even outside the mathematical domain, present little opportunity for uncertainty. If we boil water on a gas cooker the time taken will vary depending on the heat source, but with an adequate heat source the water will boil. As domains change the situation becomes more complex. If we take vaccine statistics as an example, there seem to be high enough percentages of people who avoid going to hospital with various vaccines but it isn't 100%, it's in the 80% range. So there is some slippage or a gap between what we expect and experience. Avoiding uncertainty is efficient cognitively, and Daniel Kahneman (2012) called it *fast or emotional system one thinking* that contrasted with *slower deliberative system two thinking*. Fast thinking uses heuristics (shortcuts) and while these cognitive strategies used to be considered as error prone, more recent work reviewed by Ralph Hertwig and Thorsen Pachur (2015) has shown them to be effective and efficient in certain conditions. Dewey's (1933) account of reflective thinking is closely aligned with system two thinking. Modelling thinking is central to constructivism and in this paper I outline my use of Maturana's insights into how we manage cognitive uncertainty relying especially on his paper on searching for a compelling argument (1988). My initial work involved seeking to challenge childrens' prejudices by questioning and using counter-examples rather than more direct approaches such as telling them to change their incorrect prejudices.

My discovery of Maturana's work arose while engaged in a European research project designed to promote equality of opportunity for boys and girls in schools (Drudy, 1991). The European project implemented in Ireland was designed to change teacher attitudes or behaviour in relation to boys' and girls' education in schools with a view to promoting equity. At the time computers, science and mathematics were often thought to be topics more suitable for boys. Now this issue is referred to as STEM (science, technology, engineering and mathematics) and continues to be examined as trends change slowly (Schmuck, 2017). At the time in the late 1980s as part of the Irish initiative, I proposed a constructivist program designed to facilitate change of gender stereotypes held by primary school children (Gash, 1991). I took a Piagetian approach and encouraged participating teachers to teach classes that involved questioning gender stereotypes and proposing alternatives. The central idea was to provide children with an opportunity to reflect on and reconsider their (emerging) views on the work men and women did in the workplace and how men and women differed in terms of personal-social attributes. In practice some teachers introduced male nurses and female soldiers and policewomen to children in their classrooms. Other teachers discussed women who were famous historically. Maturana's (1988) ideas about objectivities in parenthesis are about how different people can have different ways of thinking about something. They acknowledge that these different ways of thinking may each have their own legitimacy. Discussion of these ideas is needed to check them and such exploration acknowledges the other person's right to think differently. There are usually more ways to go from A to B and some of them may be equally valid. Objectivities were central to proposals concerning how teachers could discuss different constructions of gender without being confrontational. In fact the students were anxious not to be seen to be preaching or imposing their views on gender. In the case of children who have learning difficulties or who are from other countries my perception is that the students are overwhelmingly positive. In this sense the teaching approaches were rather neutral, whatever the personal views of the student teachers.

Maturana's Influence on My Research

Following Maturana's presentations in Dublin I began to use his ideas (Maturana, 1988) in papers on tolerance focusing on parenthesizing objectivities (Gash, 1992a, 1992b, 1993). Here was a striking way of giving a certain type of legitimacy to different constructions and views, whether it was thinking about gender stereotypes, or learning difficulties, or children from other cultures (Gash & Murphy, 2004). The different constructions in turn imply recognizing the importance of cognitive processes in our understandings. It follows that the assumptions made in reasoning are important in understanding how people come to hold their views. Maturana described social phenomena as requiring biological love: "the emotion that specifies the domain of actions in which living systems co-ordinate their actions in a manner that entails mutual acceptance" (1988, pp. 64–65). Social relations and interactions require mutual acceptance in the phrase *legitimate other*. This highlights the underlying ethical implications of accepting responsibility for our own reasoning. In the face of differences in thinking about an issue, recognition of our own responsibility for our own constructions suggests the need for a discussion between those holding different views whom we consider legitimate others (Gash, 2000).

The ethical implications of constructivist theory were discussed in some detail in a paper on ethics and gender presented at a conference on gender and peace at the University of Limerick in 1991 (Gash, 1993). The historical context to the paper was work suggesting that men tend to prefer hierarchical relationships and women cooperative (Tannen, 1990). My response was in part a move to avoid stereotyping and also to invoke Gregory Bateson's (1972) work on cooperative and competitive patterns of relationship and to link them to Maturana's objectivities. Cooperative relationships invite consideration of objectivities in parenthesis in which participants can consider each other's positions and discuss their origins and implications in an atmosphere that permits change. Competitive relations in contrast, eschew alternative positions and participants seek to describe the alternative view as wrong, illegitimate and not worthy of consideration in an atmosphere that makes change very difficult or impossible. This is related to the concept of viability in that knowledge in a social system must be consistent or potentially consistent to group members. Maturana (1988, p. 64) expressed this in terms of social systems depending on acceptance: "Accordingly, a human social system is defined as such by the mutual acceptance of its components in their condition of human beings." Without such acceptance, there is division and we have incompatible separate social systems.

Associated with these ideas is the concept of a consensual community that accepts a common approach to knowledge. The scientific community would be an example, religious groups would be another. Maturana juxtaposed objectivities in parenthesis with objectivity without parenthesis; the latter view does not recognize the role of cognition and its limits in knowing reality. Introducing Maturana's ideas on objectivity in parenthesis in the context of a peace conference provides a tool to discuss tolerance and to frame different approaches. In a discussion, so long as one of the participants accepts and understands constructivist ideas and how to unpack differences, then the discussion operates with mutual respect. Without mutual respect the social group breaks into us and them with the associated specters of inclusion and exclusion. There are various pitfalls including the limits of tolerance, deep disagreement, heuristics, faith, and so on. While a case can be made that each of these pitfalls are related, my thinking about these domains prompted me to think about differences that challenged tolerance.

I see constructivist theories as providing an account of the importance of the personal construction of their beliefs and thoughts. This takes place in a social context in which individual choice plays a key role. We are responsible for our constructions (Maturana, 1988, p. 31–32) and when disagreement arises I appreciated Maturana's comment about this providing an "invitation to a responsible reflection of coexistence, and not an irresponsible negation of the other" (p. 33). Change in any person's views may take time. In my own case I was initially uncomfortable about same sex marriage on the grounds that this was not my concept of marriage. I have changed my concept of marriage. I do not think tolerance is appropriate towards activities and values that damage others.

When vaccines became available, I had difficulty being tolerant of anti-vax positions. Now, it is easier for me in some cases, for example, when the person fears secondary effects. I listen, but I try to base my position on science. In the first case my tolerance was reluctant but in the second it is a form of understanding. In one of his sessions in Dublin Maturana was asked about what happens when one fails to reach agreement after a discussion in which each party has given careful consideration of the origins of their opposing views. His response has stayed with me since then: "They kill each other responsibly if they do so because they like each other and NOT because they are wrong" (Maturana, 1990). I suspect he was being provocative, but I wasn't sure! *Exclude* would suit me better. But I take from the comment another idea, that neutrality is about considering each other's views responsibly and carefully and not about unconditional acceptance and tolerance for any view the other holds.

Limits to Tolerance

Understanding the origins of different world views offers some insights into how one might discuss opposing views. However, the hope for a tolerant discussion depends on the importance of the views to the participants. In particular, views that are part of a participant's identity may be difficult to discuss with one or both participants unwilling to change views they hold firmly if not holding as objective. Amin Maalouf (1998) has written tellingly about how cultural identifies can be central in conflict when different groups in a society have different opportunities and status. This raises the question of what to do with difference in contexts when the difference is deeply felt and leads to problems. Maturana (1991) described tolerance as suspension of negation, however there are times when another person's views or another groups' views are not tolerated, for example when the views offend values and beliefs of others. In such cases context is critical and I am now aware that the degree to which some ideas essential to an individual's or group's identity are highly emotionally charged across domains such as personal social characteristics of men and women, and political and religious beliefs.

A recent study on tolerance by Hjerm, Eger, Bohman, and Fors Connolly (2020) identified three attitudes to difference: acceptance of difference, respect for difference and appreciation of difference. If tolerance is understood simply as acceptance of difference it can be seen as being neutral about differences, for example accepting variations of cultural expression or cultural dress. Forms of expression that differ may not cause offence in conditions of love (see chapters by Flavio Mesquita da Silva and Seiichi Imoto, this issue). However, other forms of difference may cause deep disagreement that can cause negative attitudes and exclusion. Unless there is discussion in a dispute that enables change such attitudes will remain. In the same vein Bartlett (2021) says that what is needed for change is the opportunity to find common ground between opposing sides. Further, when these differences are associated with group membership then holding these views can be a self-enhancing feature as in the case of populist movements where, indeed, opposition itself is self-enhancing (Gash, 2020). Hjerm et al. (2020) reported that respecting and appreciating difference is what counts in diminishing prejudice. In cases of deep disagreement this may not be possible (Fogelin, 2003).

Chris Ranalli (2018) gives an example based on creationism in which either the world is deemed to have begun according to the Bible or according to the fossil record. At issue is the epistemological basis of the opposing view that the proponents or opponents hold. In this example, science and religion are in conflict with different views held on different assumptions and with different epistemic principles. In one case the Bible is taken as providing incontrovertible evidence, in the other it is carbon dating and other sophisticated techniques including geological analysis of sediments and rocks. I found it helpful to think about these types of difference using the metaphor of different geometries based on different assumptions (Gash, 2020).

Deep Disagreement and Heuristics

I have recently worked with Spanish colleagues on deep disagreement (Fogelin, 2003). It is usually assumed that arguments follow rational processes. However, Fogelin discussed arguments without shared assumptions involving clashes of paradigms, principles and models of thinking. This debate is an exploration of the structure of the epistemological principles that underpin the metaphor of different geometries. Fogelin argued that in certain cases it makes no sense to continue arguing because logic and reason do not apply. In such cases the tensions associated with the different objectivities oblige their proponents to consider carefully the basis of their positions. Do logic and reason not apply or is there some other reason so that mutual understanding can develop using the idea of different geometries, for example?

Climate change is a topic of immediate international importance. Andrew Rivkin (2018) describes how two principles or values play a determining role in discussions on this topic, economic development and conservation of the planet's temperature. I argued that in discussions about sustainability the idea of objectivities in parenthesis might offer a way forward by inviting clarification of the bases and values of the different views taken (Gash, 2020). Reusswig (2020) challenged my optimism in a commentary to my paper arguing that it was not enough in a discussion to clarify the bases of difference and to take an epistemologically neutral position hoping that a political compromise might emerge. Rather, it was important to evaluate the quality of the opposing scientific arguments and he exhorted constructivists to think along these lines:

> It would be a magnificent task for RC scholars to find ways of qualifying and weighing competing arguments, including those of science, in order to find viable—and fast—solutions for sustainability beyond naïve realism. (Reusswig, 2020, p. 31)

In this chapter I mentioned above that heuristics plays a role in reasoning especially when predicting uncertain events. People take shortcuts and various erroneous heuristic forms of judgment have been identified by Kahneman (2012) and others. It seems that when alternative judgments are not identitarian or made in conditions of threat that compromise between opposing views is possible or easier. However, when values and identity come to the fore, then the disagreements may be deep and without easy resolution.

Part of the problem is that when it is difficult to make judgments, we may rely on our social group, and their testimony. Maturana's ideas on consensual communities (1988 p. 34) fit well here. We are familiar with faith communities, anti-vax communities and entrenched political positions. Group processes also play an important role in populism with a desire for simple answers to complex questions emphasising the will of the people and the rejection of the scientific view of reality that is regarded and rejected as elitist (Mudde & Rovira Kaltwasser, 2012). As Reusswig (2020) implied, we need to find ways to prioritize the evidence rather than be deafened by social media. I might add, in the hope of providing persuasive evidence in a context of mutual respect.

The social context is also important to how issues are interpreted and may be helpful in understanding how to create the conditions necessary for dialogue between positions showing deep disagreement (Flavio Mesquita da Silva, this issue). Social contexts change over time. In 1991 when I was engaged in research investigating how to change children's gender stereotypes, the cultural climate was such that schools accepted this intervention. I encountered little resistance from teachers or parents, nor did I expect resistance. However, in 2014 the French education minister Vincent Peillon was in conflict with parents with a similar curricular project to promote equality for boys and girls. Joseph Bamat (2014) discussed that the reason was because this time the intervention was interpreted by some as promoting the idea that children are not born as boys and girls but choose to become one or the other. Awareness of the social nature of gender has spread with different groups taking positions and the debate is not marked for tolerance. Kathleen Stock (2021) who wrote about these issues maintains that gender identity does not outweigh biological sex when it comes to law and policy. She resigned her university position in October 2021 due to pressure from students and lack of support from the university for her views (Adams, 2021; Moorhead, 2021). One of my friends has a teenage granddaughter in Paris who objects to her grandmother using the nouns *boy* and *girl*. Human rights organizations provide legal guidance to protect individuals who want to choose their gender. However, in athletics internationally there are tensions about the identity of a number of athletes with high testosterone levels who want to compete as women. The tension between inclusion and fairness has been highlighted in such cases (O'Riordan, 2021). This is an issue where identity is central. Can we advise how to promote debate and communication and Maturana's biological love in ways that conserve mutual respect? Melanie Killen, Adam Rutland and Mark Ruck (2011) summarize research on discrimination whether based on gender, age, race, ethnicity, religion, indigenous background, or other reasons. They conclude that childhood is the time to promote tolerance, equity and justice. Sadly, it seems in today's world it is difficult to conserve mutual respect in debates on many subjects including gender identity, the need to vaccinate, politics and religion.

Solutions or New Directions

In my past I found Maturana's ideas about objectivities a convincing way to try to bridge differences between different personal positions. Later, Maturana's ideas about consensual communities with their implicit acknowledgement of paradigms of knowing impressed on me the importance of community and the corollaries of inclusion and exclusion. The following ideas seem worthwhile:

- The need to promote understanding of paradigms in cognition, or an educational sense of the need to value different ways of knowing.

- The need to emphasise the importance of reflexivity concerning one's positions
- The need to advocate an evaluation of the implications of different objectivities following Ruesswig's (2020) comments above.
- The need to promote civic engagement as suggested by Lonnie Sherrod (2011). This emphasis would seem to fit well with the program implied by Seiichi Imoto in relation to promoting human rights (this issue, pp. 55–61).

References

Adams, R. (2021, 28 October). Sussex professor resigns after gender rights row. *The Guardian.* Retrieved on 22 February 2022 from https://www.theguardian.com/world/2021/oct/28/sussex-professor-kathleen-stock-resigns-after-transgender-rights-row

Bamat, J. (2014, 29 January). French parents boycott schools over gender theory scare. *France 24.* Retrieved on 22 February 2022 from https://www.france24.com/en/20140129-france-sex-education-gender-discrimination-protest-school.

Bartlett, M. (2021, 19 November). How do you argue with anti-vaxxers who believe they're on a noble mission? *The Guardian.* Retrieved 11 March 2022 from https://www.theguardian.com/commentisfree/2021/nov/20/how-do-you-argue-with-anti-vaxxers-who-believe-theyre-on-a-noble-mission

Bateson, G. (1972). *Steps to an ecology of mind.* New York: Ballantine.

Chomsky N. (1957). *Syntactic structures.* The Hague: Mouton.

Dewey, J. (1960). *The quest for certainty.* New York: Capricorn. (Originally published 1928)

Dewey, J. (1971). *How we think.* Chicago: Henry Regnery Company. (Originally published 1933)

Drudy, S., Gash, H., Lynch, K., Lavin, P., Moles, R., Lane, C., Ganly, M., Forgarty, C., O'Flynn, G., & Dunne, A. (1991). Integrating equal opportunities in the curriculum of teacher education 1988-1991: TENET Programme Dissemination Phase. *Irish Educational Studies, 10*(1), 271–289. Available at https://cepa.info/2180

Fogelin, R. (2003). *Walking the tightrope of reason: The precarious life of a rational animal.* Oxford, UK. Oxford University Press.

Gash, H. (1974). The constructivist epistemology in John Dewey, Jean Piaget, and cognitive developmental psychology. In C. D. Smock & E. von Glasersfeld (Eds.), Epistemology and education (pp. 27–44). Athens, GA: Follow Through Publications. Available at https://cepa.info/2181

Gash, H. (1991). Gender perceptions: A constructivist approach to stereotyping. *Oideas, 37,* 57–74.

Gash, H. (1992a). Reducing prejudice. *Reach, 6,* 48–53.

Gash, H. (1992b). Reducing prejudice: Constructivist considerations for special education. *European Journal of Special Needs Education, 7,* 146–155.

Gash, H. (1993). Constructivism, ethics and gender: Implications for education. *Irish Educational Studies, 12,* 227–240.

Gash, H. (2000). Epistemological origins of ethics. In L. P. Steffe & P. W. Thompson (Eds.), Radical constructivism in action: Building on the pioneering work of Ernst von Glasersfeld (pp. 80–90). London: Falmer.

Gash, H., & Murphy, E. (2004). Children's perceptions of other cultures. In J. Deegan, D. Devine, & A. Lodge (Eds.), *Primary voices: Equality, diversity and childhood in Irish primary schools* (pp. 205–221). Dublin: IPA.

Gash, H. (2020). Constructivism, fast thinking, heuristics and sustainable development. *Constructivist Foundations, 16*(1), 1–12.

Glasersfeld E. von (1974). Piaget and the radical constructivist epistemology. In C. D. Smock & E. von Glasersfeld (Eds.), Epistemology and education (pp. 1–24). Athens, GA: Follow Through Publications. Available at http://www.vonglasersfeld.com/034

Hertwig, R., & Pachur, T. (2015). History of heuristics. In J. D. Wright (Ed.), *International encyclopedia of the social and behavioural sciences, Volume 10.* (2nd ed., pp. 829–835). Amsterdam: Elsevier.

Hjerm, M., Eger, M. A., Bohman, A., & Fors Connolly, F. (2020). A new approach to the study of tolerance: Conceptualizing and measuring acceptance, respect, and appreciation of difference. *Social Indicators Research, 147,* 897–919. https://doi.org/10.1007/s11205-019-02176-y

Killen, M., Rutland, A., & Ruck, M. (2011). Promoting equity, tolerance, and justice: Policy implications. *Sharing Child and Youth Development Knowledge, 25,* 1–33.

Kahneman D. (2012). *Thinking, fast and slow.* London: Penguin.

Maalouf A. (1998). *Les identités meurtrières.* Paris: Grasset.

Maturana, H. (1988). Reality: The search for objectivity or the quest for a compelling argument. *Irish Journal of Psychology, 9,* 25-82. (Special issue: Radical constructivism, autopoiesis and psychotherapy. Ed. Vincent Kenny.)

Maturana, H. (1990). Seminar given by Humberto Maturana to psychologists. Dublin, 1 December 1990.

Maturana, H. (1991). Science and daily life: The ontology of scientific explanations. In F. Steier (Ed.) *Research and reflexivity* (pp. 98–122). London: Sage.

Moorhead, J. (2021, 22 May). Kathleen Stock: Taboo around gender identity has chilling effect on academics. *The Guardian*. Retrieved on 22 February 2022 from https://www.theguardian.com/education/2021/may/22/kathleen-stock-taboo-around-gender-identity-chilling-effect-on-academics

Moscati, I. (2016). Measuring the economizing mind in the 1940s and 1950s: The Mosteller-Nogee and Davidson-Suppes-Siegel experiments to measure the utility of money. *History of Political Economy, 48*(Supplement 1), 239–269.

Mudde C., & Rovira Kaltwasser, C. (2012). *Populism in Europe and the Americas: Threat or corrective for democracy?* Cambridge: Cambridge University Press.

O'Riordan, I. (2021, 27 November). Ian O'Riordan: No running from DSD and transgender debate in athletics. *The Irish Times*.

Piaget, J. (1954). *The construction of reality in the child*. New York: Basic Books.

Piaget J. (1970). Piaget's theory. In P. H. Mussen (Ed.), *Carmichael's manual of child psychology* (pp. 703–732). New York: Wiley.

Ranalli, C. (2020). Deep disagreement and hinge epistemology. *Synthese, 197*, 4975–5007.

Reusswig F. A. (2020) De- and re-constructing sustainable development. *Constructivist Foundations, 16*(1), 030–032. Available at https://constructivist.info/16/1/030

Rivkin, A. (2018, July). Climate change first became news 30 years ago. Why haven't we fixed it? *National Geographic*. Retrieved on 22 February 2022 from https://www.nationalgeographic.com/magazine/article/embark-essay-climate-change-pollution-revkin

Schmuck, C. (2017). *Woman in STEM disciplines*. Cham, Switzerland: Springer.

Sherrod, L. (2011). Commentary: Equality and justice in developmental science. in M. Killen, A. Rutland, & M. Ruck, Promoting equity, tolerance, and justice: Policy implications. *Sharing Child and Youth Development Knowledge, 25*, 31–32.

Skinner, B. F. (1957). *Verbal behavior*. La Jolla, CA: Copley Publishing Group.

Stock, K. (2021). *Material girls: Why reality matters for feminism*. Little Brown.

Tannen, D. (1990). *You just don't understand: Talk between the sexes*. Morrow.

Bunnell, P. (2021). *Organization & Structure*. Photograph.

How My Understanding of Languaging in Non-Speaking People With Autism Has Been Informed by Conversations With Humberto Maturana

Kathleen Forsythe[1]

> This paper explores current areas of my work that have been deeply influenced by Maturana's thought as well as my work and conversations with Dr. Pille Bunnell over the past 25 years. I live and work with autistic young people who are non-speaking. Given that Maturana has posited that all words we use orient our sensory-operational-relational living in the flow of emotions, giving rise to what we know and share in our language interactions, I have wondered at the knowledge architecture of those who do not speak. I am observing learning in those who appear not to use names and words, yet who develop their understanding using different strategies according to the developmental trajectory of their sensory motor system. This paper queries whether non-speaking autistic people live dominantly in the non-languaging, three-dimensional present of Maturana's *zero time*, which leads them to experience languaging in a no less valid but very different domain of communication.
>
> **Keywords:** dynamic temporal architecture, languaging, emotioning, autistic, observing for learning, neuroception, social engagement, zero time

Introduction

I was fortunate between the mid 1980s and the late 1990s to have had a number of opportunities where conversations and experiences with Humberto Maturana helped me to deepen in my understanding of Humberto's thinking in the biology of love and the biology of cognition.

The practical application of these ideas in my work has included the concept of observing for learning as an alternative form of assessment (Forsythe, 1990), the development of educational systems grounded in observing for learning and co-inspiration;[2] and the development of organizations based on love as the only emotion that expands intelligence. I have lived these ideas in my work as an educational leader over the past 35 years.

Lately I have been living and working with young people with autism who are non-speaking. As someone who is curious about how we know and communicate through conversation, coming to live with those who, on the surface, seem unable to enter conversation, has helped me to better articulate the dynamic epigenic[3] architecture of how we humans come to make sense of our experiences and how we share meaning with each other. This paper explores current areas of my work that have

---

1. Doctor of Knowledge Architecture, New Dawn Education & Research Company Ltd.
   Email: kathleenforsythe@mac.com
2. Website: www.selfdesign.org
3. development on top of prior development, as in hysteresis

been deeply influenced by Maturana's thought as well as my work and conversations with Pille Bunnell over the past 25 years.

The Dynamic Nature of Knowing

To begin I will outline my current thinking on the dynamic nature of knowing. At a conference in Belo Horizonte in November 1997 Humberto Maturana was asked about causality. He leapt up and exclaimed: "There is no causality. There is a dynamic temporal architecture" (personal notes from the experience).

This idea of the dynamic temporal architecture electrified my thinking and was one of the ways I came to understand how it is through the local linear where we may observe a connection of dependence or precedence while at the global systemic level we cannot account for everything that happens that gives rise to a particular moment that I distinguish in my observing.

Building on this notion of a dynamic temporal architecture I came to the idea that a dynamic epigenic architecture speaks to how it is we humans know. Knowledge does not exist independently of the humans who dynamically generate it through their interactions with themselves and others. And it is through languaging that we come to share our unique knowledge architectures, generating the extrasomatic knowledge of our species and our cultures through engaging with each other. We then act as if this "knowledge" exists as "reality" independent of the human generative system of languaging that both composes and maintains the recurrent patterns of our experiences.

Here is how I have been thinking about languaging and knowing, deeply grounded in Maturana's thinking:

- Languaging requires relationships. It is the coordinating of coordinated actions through which observers make meaning.
- Maturana said emotions are domains of relational behavior such that we are in effect different being capable of different activities according to our emotions. This I summarize as emotions are dispositions for actions … they work in the domains of neuroception,[4] perception and conceptualization. They reflect changing body states.
- Body states shift constantly in response to both internal and external perturbations of our sensory effector systems.
- Body states shift in a coordinating flow with the shifting flow of the body states of those we interact with, as exemplified by the importance of early mother child relationships in the development of languaging.
- From this coordinating of body states experienced neuroceptively and perceptually,
- We begin to name what we distinguish in relationship. "What we distinguish

---

4. Neuroception is a word coined by Dr. Stephen Porges that refers to our unconscious ability to assess for risk, danger and safety, often outside the realm of our awareness. https://www.stephenporges.com/

is brought to existence as we distinguish it with what we do and name" (Maturana, 2014, p. 188). Through abstraction of the experiences we name we begin to conceptualize. In this manner meaning arises when we share the naming with others. This can allow us to generate a domain of explanation independent of the actual experience. This is how psychic and mental states emerge through the coordinating coherences of these coordinating experiences.

- The braiding of the emotioning—embodied states—and languaging generates a domain of shared coherences of experience that although originally generated from embodied states, can begin to seem as if it exists independently of the perceived moment. This is how we begin to know a "reality" and then live in it as if it were independent of our actualization as living systems in an emergent wavefront structurally coupled to our environment, internally and externally to our physiology.
- We all generate many domains in which we live and share experiences. Our domains are unique to us as we generate them from our history and lineage of experiences. They overlap with others when we share a common history of a particular kind of braiding of emotioning and languaging in our cultures.
- "We put names to what we distinguish, but since we do not distinguish independent entities, but distinguish sensory effectors configuration in our living, what we name are sensory effectors configurations that pertain to the coherences of our living" (Maturana, 2014, p. 187). From these configurations of coherences, an epigenic structure is generated by the coherences of experience among humans and distinguished in language.
- The sensory effector configurations that pertain to the coherences of our living form a dynamic epigenic architecture where meaning arises through shared interaction … and each person contributes their unique knowledge architecture to the generative knowledge we share as a species.

Given all the above, I am led to ask what is the basis of the coherence of experiences such that meaning arises in the dynamic epigenic architecture? My answer is: our shared physiological system, namely the sensory motor effectors which serve to structurally couple us to the emerging environment. Our common environment provides the perturbations that begin to cohere in patterns that eventually arise as the recursively coordinating coherences of experience known as languaging.

A mother coos at her baby, smiling with her eyes and her mouth. She rocks the baby in her arms. Her voice prosody and crinkly eyes and facial expression neuroceptively trigger feelings of safety in the baby's body and the baby relaxes from the stressors of internal sensations and falls asleep. Eventually after countless such interactions the baby's system learns to recognize the source of comfort and to seek the patterns that settle the body. This can involve a reciprocal interaction where the baby's eyes and mouth change in correlation with the mother's and we say "Oh, see the baby is smiling." Eventually the reciprocity extends to body movements, eye contact, facial expression, gestures and vocalizations, all of which form the foundations of languaging.

As Maturana describes this:

> Although we do not usually see it in this way, we live immersed, so to say, in a flowing dynamic network of changing sensations in which from the moment we are conceived, we learned to abstract the sensory configurations that begin to guide the course of our living according to the manner of living that we learn-generate-create as we live. And in this network of sensations, what we distinguish is brought to existence as we distinguish it with what we do and name, much as a child in a sandy humid beach brings forth stars, triangles, flowers ... with the moulds that he or she may happen to be playing with. So, names and words in general are not trivial artifices for indicating pre existing conceptual or physical entities, they connote what we do and feel as we use them. Without our always being aware of what we are doing, names and, in fact, all words that we use, constantly orient our sensory-operational-relational living, both illuminating and obscuring it, according to the emotions. (Maturana, 2014, p.188)

When a child does not respond in a manner that we have come to expect of our infants and toddlers, then we may become concerned and begin to seek explanations. Maybe the child does not make eye contact or smile or respond to our voice. Maybe the child does not speak. We begin to wonder, "What must have happened such that he or she acts in this manner?"

Languaging arises within an environment of other humans who function as caring observers responding to the perceived needs of the baby (Maturana, Verden-Zöller, 2008). Our long childhood leads to coherence within the cultural domain of caregiver and child. Children raised with wolves and not humans, for example, may not fully realize their full human capacity for languaging, when their dynamic epigenic architecture takes a path of epigenesis that is not coherent with others of their kind (Maturana 1987).

So what happens in autism such that some children do not develop language? What happens in the braiding of emotioning and languaging such that some with autism have difficulty having any conversation? What happens for those autistic persons who do not speak at all? In what way are they living in languaging?

Some Background

Lakoff and Johnson have explored how our conceptualization system is anchored in physical orientation as exemplified by our use of metaphor and analogy. They further explain that it is the embodiment that allows for dynamism and generalization in language (Lakoff & Johnson, 1999). It is in this embodiment that we often see the challenges of *autists* who do not develop languaging as their physiological states differ from those held by the already languaging members of our species.

In our living we encounter multiple domains of experience and, through our interaction with others in languaging, multiple domains of conversation. This enables us to conceptualize our experiences. Often we live in these conceptualized domains which can obscure the original sensory-operational-relational living and the body state shifts that gave rise to our distinctions in the first place. We can get lost in a novel or concentrate with work and forget our body experiences. This is the case for all of us as we begin to live in our minds as if we were disembodied! Those with autism may not be able to correlate their experience when engaging with another person in order to co-generate coherence from shared conceptual domains. The link between the domains of their sensory-operational-relational living and languaging may become so tenuous that they are unable to have their body realize what they are thinking.

This can lead to the lack of generalization in conceptualization and perception across people and circumstances, a common challenge of those with autism, especially those who are non-speakers. It may lead to difficulties in social engagement and conversation in those autists who do have language but often seem unable to listen to what the other person is saying or to discern whether the other is listening to them or not.

Autism may be an example of a variation in the coupling of body state changes with experience resulting in incoherence rather than sensory effector coherence. In some cases it may mean a rigid lock between some particular body state and experience such that responsiveness to perturbations that would be meaningful for understanding is lost. If the link between the experience and the conceptualization is tenuous, as may be the case with many non-speaking people with autism, then language is dominantly perceived as sound, not conceived as meaning. For example, M, the young non-speaking man I live with, could sing songs based on the sounds of the words. When he learned to read the words, he could then sing the words rather than the sounds. His original coherence was with the sound and not the articulation or meaning of the song. We did not understand this until he was able to read and sing in his early twenties. His understanding of singing was acted on showing that there was understanding yet it masked that his understanding did not include conceptualization that enabled abstraction. Like many people, children with autism require concrete experiences to grasp abstract concepts. Yet the variation in their sensory effector correlations may lead them to different pathways of understanding. This is most noticeable in those who are non-speaking.

Observing for Learning

I live, in my daily life, with two non-speaking young adults who have profound autism. Much is made of what a person with autism cannot do. Observing for learning[5] is providing me with an expanded understanding of the extraordinary way they do make sense of their sensory-operational-relational living. A small example of what I am observing is the emerging of threeness.

---

5. Observing for learning is an assessment practice that I developed in 1990, which is grounded in Maturana's notion that "Everything is said by an observer." In observing another in order to understand their learning one has to be aware of the domains from which one is observing in order to be able to see the validity of the other's experience. In this sense learning can look many different ways.

K is an autistic 22-year-old who is non-speaking. She can act as if she is a young child of two or three. Yet she is also an aware young woman with an extraordinary grasp of what is going on around her. She can read and she can spell even though her dyspraxia prevents her from signing, speaking or typing spontaneously. K is now able to play the piano, guitar, drums, recorder and xylophone and to read music at an elementary level.

Through observing her learning, I have come to really appreciate the distinction between perceptual and conceptual learning. K's unusual sensory development means that her conceptualization has developed differently as well. K learns perceptually and her flow between the perceptual and conceptual dynamic can be attenuated. For example, now that she reads notes, her teacher presents the following set of notes to her randomly and she is asked to demonstrate the number of beats she sees by playing or striking the instrument.

Figure 1: Note Sequences as Presented to K.

Her response is often accurate and she counts the number of notes (e.g., 1, 2, 3 or 4 notes). However, she also repeats the sequences so that, if she is shown 4, then 3 she will often do the 4 again, relying on her visual pattern memory and not responding to the second image as presented. If she is prompted to look again and the support person's finger goes beneath each note, she always does these accurately. Recently she looked at the 3 note sequence, said "3," and proceeded to hit the instrument 3 times without prompting.

This helped me to see that for her, each time the pattern is presented, she does not necessarily see groupings of 1, 2, 3 and 4 that translate into the numerical concepts but must reconstruct the number sequence, with support, each time. She does this from memory as well as from counting. However, her occasional use of her voice to say *three* indicates that the conceptual notion of three is present and that when she is able to connect to this conceptual framework, she overrides her perceptual strategy for making sense of the request. This gives us a glimpse of the strategies she has developed to make sense of her experiences when she has been unable to use language.

As Maturana has said:

> Without our always being aware of what we are doing, names and, in fact, all words that we use, constantly orient our sensory-operational-relational living, both illuminating and obscuring it, according to the emotions that they evoke in us. (Maturana, 2014, p. 188)

Observing K who is someone who does not use words, has helped me to see how, for her, she does not necessarily orient her sensory operational relational living within domains that intersect with those with whom she lives.

In coming to better understand how K understands so that we may better share our experiences, Maturana's ideas have informed much of my observing for learning.

- If the braid of emotioning and languaging is uneven, tenuous or masked, then the dynamic epigenic architecture develops into multiple domains that are not always intersecting or in recurrent interaction. K's number concepts linked to the symbols *1, 2, 3, 4*, and comprehension of language linked to the sounds *one, two, three, four* exist but the domains in which they arise for her may not always intersect.
- Time as we know it may not be perceived in the same way by those who do not develop languaging in the usual way. hence the need to anchor sequences and patterns in daily living to better navigate the sense of a world unfolding in a changing present without the capacity to impose a conceptual pattern on the arising. For K, the sequence in which something is presented may determine her response. She may not be able to generalize an idea that is presented out of sequence.
- Those of us who live in languaging also live in the cultural invention of time. Although we also live in a dynamic temporal architecture, once we are living in language it is hard for us to return to awareness of the changing present and to understand that past and future are ways of speaking about what is essentially Maturana's ideas of *zero time* or *no-time*. Maturana proposed time as "an imaginary new dimension of space in the three-dimensional present in which our non-languaging existence occurs" (Maturana, 2008, p. 3). This leads me to wonder whether those with autism who are non-speaking and hence do not live in languaging live primarily in Maturana's no-time, "the three-dimensional present in which our non-languaging existence occurs" (Maturana, 2008, p. 3).
- Languaging, by its very generative nature, puts us in time as an imaginary dimension of space. We regularly conceptualize and unconsciously project schedules on our daily living. We can think to ourselves that the lesson will be over in a few minutes or at 10:30. Without conceptualizing time in languaging, a person who makes sense of the world through sequences, patterns and visual cues seeks anchors to let them know where they are in the unfolding process of a non-languaging existence.
- K will sometimes answer the request for the note sequence, not from what she sees in front of her but from what she remembers of the last sequence. These two domains do not appear to intersect. For her they may not be two domains, just different responses in the present, without any real way of selecting one or the other. In the same sense if any element of the music lesson has been missed or forgotten, she will exhibit high anxiety until we are able to complete it. This is often interpreted as the autistic person being obsessive when it may be a strategy developed to make sense of a lesson that unfolds in the changing present without always linking to the conceptualized timeframe

that is inherent in languaging. If a step is missed in the sequence—e.g., A should have happened after B, so it's all wrong if A is missing.
- K's experience of our music lesson may be quite different than mine but it is no less valid. My role in providing support is to imagine the domain in which her reactions are valid and to both interpret and adjust the flow of the lesson to support the non speaking person's way of understanding while at the same time providing the safety and security for her to more easily flow with changing circumstances. This is the trueness of Maturana's statement: "Love is the only emotion that expands intelligence." For K to arise as a legitimate other in co-existence with myself, in love, even though her actions are not what I might expect, it is important for me to better understand her way of experiencing the world (Maturana & Verden-Zöller, 2008).
- For threeness to emerge for K, as a concept in language, she must experience it many times in many different circumstances for it to override her perceptual understanding of threeness (counted on the fingers visually, tactiley and kinesthetically) and enable her to respond to 3 fluidly and conceptually through languaging. Observing her learning these distinctions, reveals how challenging it is to bring her motor skills into alignment with her perceptual and conceptual understanding and how many of her domains of experience do not appear to intersect as she is not connecting these separate events semantically in language.

Dr. Stephen Porges has also shown that some with autism have had damage to the stapedius muscle in the middle ear that has affected their capacity to distinguish the frequencies of human speech (Porges et.al., 2013). When this is improved, through his Safe & Sound Protocol, significant changes in understanding are observed. This was indeed the case for K who spent most of her life until the age of 18 trying to make sense of sensory-operational-relational living, without the benefit of the words used by others to orient and distinguish experience.

For the years in which she could not easily distinguish words, she developed strategies for compensating. She appeared to use vision and memory, for example, to understand what was said rather than the auditory processing of words, anchoring her understanding in a particular sequence of actions or a person or a place.

Languaging is a whole body process that includes breathwork, prosody, and body state change as well as the use of sounds as tokens of coordination and meaning as we navigate our sensory-operational-relational living. There is some suggestion that it may be in the action-oriented neurological structure that a delay occurs between the affect and motor neuron in those autistic people who are non-speaking. They may be understanding what is said and may be attempt to respond verbally but their body does not react. The motor neurons, for whatever reason, do not act on the mind's command. This can extend also to the person's capacity to print, write, sign and type.

I discussed this with Humberto Maturana the last time I met with him in 2012 at the ASC conference in Asilomar. He was very interested in my observations and agreed that it was possible that the motor neuron output signals may indeed not be functioning in tune with the afferent neurological structure. If this were indeed the case it would impair the development of conversation as a child develops. This may appear as motor dyspraxia. Motor dyspraxia, the inability to plan actions, can affect everything, from speech to personal care to learning. It is thought to be an immaturity in the neuronal development of the motor neurons.

If the motor neuron system is an area of delayed development then the idea of emotions as dispositions for action would look very different in someone whose system of action does not function in a responsive manner. Recent work on the role of the vagus nerve and autonomic nervous system, in the development of what Dr. Stephen Porges (2011) calls the *social engagement system,* supports the connection between the sensory and motor systems as being fundamental in how mammals develop both the awareness of each other as well as the capacity to determine whether it is safe to interact or not. This extends to cross-species communication such as the relationship we develop with our pets.

Autism is characterized by impairment in social engagement. Lack of social engagement is often seen early in a child's life even if not diagnosed as autism. If the coherence of sensory-operational-relational living, between the infant and caregiver does not arise due to a difference in the child's sensory motor developmental trajectory, consensual coordination between the mother and infant will also develop differently. The baby may not smile and coo back to the mother or make eye contact. The baby's body response may be seen by the caregiver as rejection and that the infant wants to be left alone.

As this trajectory is followed through development, the child does not develop the ability to further coordinate their consensual coordination and develop language. Further the braiding of languaging and emotioning that is the hallmark of conversation cannot develop. Thus for non-speaking people this can lead to a pattern of isolation that can lock them inside a world that is not recognized by others. It is only through their high anxiety and meltdowns that we see the level of their frustration at being unable to exhibit the emotions they may feel or communicate what they think. Too often these behaviors are treated as aberrant and to be extinguished, without any sense of compassion for their root causes. Frequently these intelligent and sensitive people are treated as being cognitively impaired and the extraordinary strategies they use to compensate for the vagaries of their development are ignored.

Living with two non-speaking young adults I remain in awe of how they make sense of the world without living in languaging as we know it. They are diagnosed with profound autism and are unable to tell you their names or coordinates. They require round the clock care and support with all aspects of daily living.

I have come to realize that although they have fully developed minds these minds are not always in control of their actions. We use language as a control mechanism. Without language, along with the disconnect between the sensory/effector loop many non-speaking autists are really locked in their bodies. Normally our languaging nervous system develops in structural coupling with the rest of our nervous system, and this is not easily accessed by those who do not speak. The regulating function of language may be weak at the neuronal level, yet we are discovering that languaging is still possible even if unreliable.

This is often seen in the inability to indicate yes or no or even to choose left and right even though it is clearly something they have demonstrated understanding in other situations. Yet both are able to read music and play instruments, to take dance and music classes through Zoom and engage with the instructors—all of this with local support. Both can demonstrate that they can read and even though their motor dyspraxia prevents them from speaking, meaningfully signing, typing or handwriting. They participate in the daily gardening and baking work at our home where we sell the products of their work. Most importantly, the two young people I live with (aged 20 and 22 at time of writing) are beginning to enter into conversations, albeit not always in spoken words, rather in their daily living flow of consensual coordinations of consensual coordinations, as well as in limited distinguishing and naming, that is, in the first and second recursions of consensual coordination (Bunnell, 2018).

I have come to realize that these non-speaking humans indeed live in a multiverse of domains of experience that do not easily cohere with those within which I live. Yet there are still many intersecting domains of understanding such that we can live together in the flow of everyday life. Understanding how differently, yet validly, they may experience the world, has helped those who support them to observe their learning in order to understand their perspective, and thus adapt to connect meaningfully with them as well as we are able. Even though their inability to speak is a profound disability in members of a languaging species, they live rich and full lives.

I am forever grateful for the deep humanity in Humberto's thinking and for the gift I had of actually being able to be in conversation with him over many years. It has shaped my life and work for over 35 years and enabled me to contribute to better understanding and appreciating the worlds of others. Most importantly it has given me a framework for wondering about the mysterious nature of how we humans do indeed bring forth multiversa together in our desire to see and be with each other in our legitimacy—whether languaging or not!

Acknowledgements

I acknowledge the parents of K and M, Janice Terashita-Clark and Peter Clark, for their permission to use examples from their children's daily living.

I acknowledge Dr. Pille Bunnell for many years of conversation about knowing and learning grounded in Dr. Maturana's work and how it could be applied in all aspects of education.

References

Bunnell, P. (2018a). Languaging: A spark, some light ... and dark—part 1. [video]. *YouTube*. Retrieved March 15, 2022 from https://www.youtube.com/watch?v=9V0KRALjgTY

Bunnell, P. (2018b). Languaging: A spark, some light ... and dark—part 2. [video]. *YouTube*. Retrieved March 15, 2022 from https://www.youtube.com/watch?v=HqUxkMijjuM

Bunnell, P., & Forsythe, K. (1999). *Intelligence arises in relationship*. Asilomar, CA: International Society for Systems Science.

Forsythe, K. (1986). Cathedrals in the mind: The architecture of metaphor in understanding learning. In R. Trappl (Ed.), *Proceedings of the European Congress on Cybernetics and Systems Research, Cybernetics and General Systems*. Amsterdam: Reidel Publishing Co.

Forsythe, K. (1988). A disposition for wonder: The art of conceptualization in the architecture of being human. In R. Trappl (Ed.), *Systems and Cybernetics 88*. Dordrecht: Reidel Publishing.

Forsythe, K. (1990). Designing a learner profile system that teachers would want to use. Proceedings of the Seventh International Conference on Education and Technology, March 21-22, 1990, Brussels

Forsythe, K. (2018). Open peer commentary: Conserving the disposition for wonder in new horizons for second-order cybernetics. In L. H. Kauffman (Series Ed.) & A. Riegler, K. H. Müller, S. A. Umpleby, (Vol. Eds.), *Knots and everything: Vol. 60. New horizons for second-order cybernetics* (pp. 109–112). Singapore: World Scientific.

Lakoff, G. & Johnson, M. (1980). *Metaphors we live by*. The University of Chicago Press.

Lakoff, G. & Johnson, M. (1999). Philosophy in the flesh: The embodied mind and its challenge to western thought. New York: Basic Books.

Maturana, H. R. & Varela, F. J. (1987). *The tree of knowledge: The biological roots of human understanding*. Boston: Shambhala.

Maturana, H. R. (1987). Everything is said by an observer,.In W. I. Thompson (Ed.), *Gaia: A way of knowing* (pp. 65–82). New York: Lindisfarne Press.

Maturana, H. R. (1988). Reality: The search for objectivity or the quest for a compelling argument. *Irish Journal of Psychology, 9*(1), 25–82.

Maturana, H. R. (1988). Ontology of observing: The biological foundations of self-consciousness and the physical domain of existence. In R. E. Donaldson (Ed.), *Texts in cybernetic theory: An in-depth exploration of the thought of Humberto Maturana, William T. Powers, and Ernst von Glasersfeld*. American Society for Cybernetics (ASC) conference workbook. ASC conference held in Fulton, CA, October 18-23, 1988. (Originally published as: Maturana, H. R. [1987]. The biological foundations of self-consciousness and the physical domain of existence. In E. R. Caianiello [Ed.], *Physics of cognitive processes* [pp. 324–379]. Singapore: World Scientific.)

Maturana, H. R., & Nisis, S. (1997). *Human awareness: Understanding the biological basis of knowledge and love in education* (J. Cull, Ed.). Retrieved March 23, 2022 from http://members.ozemail.com.au/~jcull/articles/arteduc.htm

Maturana, H. R. & Bunnell, P. (1997). *What is wisdom and how is it learned?* Paper presented in abridged form at the North American Association for Environmental Educators conference, July 1997. Available at https://cepa.info/658

Maturana, H. R. & Poerksen, B. (2004). *From being to doing: The origins of the biology of cognition* (W. K. Köck & A. R. Köck, Trans.). Heidelberg: Carl-Auer.

Maturana, H. R. (2008). Comments on the Occasion of the 2008 Weiner Medal, American Society for Cybernetics. Retrieved March 23, 2022 from https://asc-cybernetics.org/2008/HM-08WienerComments.pdf

Maturana, H. R., & Verden-Zöller, G. (2008). *The origin of humanness in the biology of love* (P. Bunnell, Ed.). Exeter, UK: Imprint Academic.

Maturana H. R. (2014). Understanding Social Systems? *Constructivist Foundations, 9*(2),187–188.

Porges, S. W. (2011). *The polyvagal theory: Neurophysiological foundations of emotions, attachment, communication, and self-regulation*. New York: W.W. Norton.

Porges, S. W., Macellaio, M., Stanfill, S. D., McCue, K., Lewis, G. F., Harden, E. R., Handelman, M., Denver, J., Bazhenova, O. V., Heilman, K. J. (2013). Respiratory sinus arrhythmia and auditory processing in autism: Modifiable deficits of an integrated social engagement system? *International Journal of Psychophysiology, 88*, 261–270.

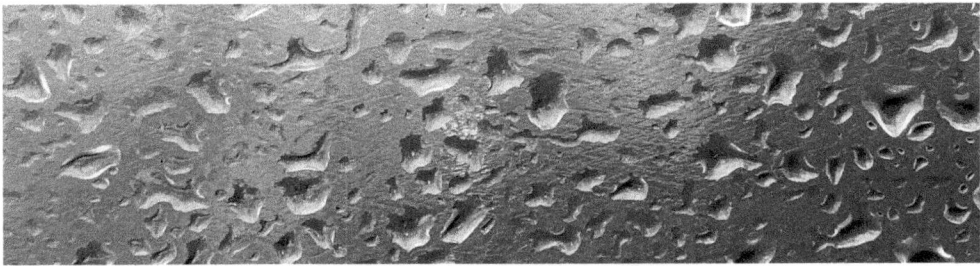

Bunnell, P. (2019). *Droplets*. Photograph.

Bunnell, P. (1995). *Maturana as Playful*. Photograph.

Consciousness: An Unresolved Question, and What Maturana Has to Say About It

Pier Luigi Luisi[1]

> Some of the main papers of Humberto Maturana dealing with consciousness are examined, with the aim of comparing his view with the main literature in the field. In a first part, a brief review is offered of the most accredited authors, broadly divided in two camps: those who maintain that consciousness is a property of the brain, and those who instead claim that consciousness is primary. This analysis shows that the comparison between Maturana's view and literature cannot be easily done; namely he cannot be allocated to either of the camps in the current literature. Maturana has his own particular view on the subject, as he assumes that consciousness is rather the result of languaging and human relationship, or, as he says "Human existence occurs in the relational and operational space of living."
> **Keywords:** consciousness, self-consciousness, cognition, autopoiesis, brain, neurobiology, nervous system, structure-determined, languaging, information, neuro-phenomenology

A Quick Look at the Literature on the Study of Consciousness

Let me first mention a landmark paper by the Australian philosopher David Chalmers (1995), who proposed to distinguish between the *easy* and the *hard* problems of consciousness. The easy one has to do with perception—for example, I see the blue sky and this seeing is a physical phenomenon, due to a signal on the retina connected somehow to the brain and the brain somehow producing the blue color. The process is not at all easy, but at least one sees the cause and effect and this is what Chalmers meant by the term *easy*. However, you can also, simultaneously, be immersed in the feeling of blue, immersed in a sensation which is a personal experience, something quite subjective. This dimension of feeling is for Chalmers, and others after him, the hard problem of consciousness. In fact, there is no organ and no part of the brain directly responsible for feelings. What is the nature, and the origin, of this feeling?

It is generally accepted that the act of consciousness is equivalent to an act of personal experience. Thus consciousness, in this sense, is a very important part of oneself, as it creates our inner world—all what happens inside you, even without your conscious desire or attention, whether it be the faraway barking of a dog or the feeling of love, fear, or anger, being surprised by spontaneous memories, getting lost in the smell of an orange (Humphrey, 2009) and so on. In simpler terms, as Bergonzi puts it, consciousness is *presence* (Bergonzi, 2011). The fact that consciousness forms your internal world, for many people, is also equivalent to saying, that there are two different descriptions of the world, the objective and the subjective. Is it so, that there

---

1. Email: luisi@mat.ethz.ch

are really two realities? This is a first important question within the realm of consciousness.

The fact that there is no physical correlate in the brain or the body for personal experience, places consciousness in a very particular ontological position. Here is what philosopher Michel Bitbol writes about this point:

> Consciousness is the name we give to the astounding realization of immediate existence, even before its more intricate connotations such as reflective self-consciousness or moral conscience. ... These obvious (yet destabilizing) remarks are not derived from any reasoning. They express what E. Husserl ... called a phenomenological description; a plain statement of what is immediately experienced, irrespective of any interpretation of the contents of experience in naturalistic terms. (Bitbol, 2011, p. 1)

And, on the same page in the same article:

> Any question about consciousness is then utterly philosophical. For when we raise a question about consciousness, we are not only implicated in it in abstracto, timelessly, as generic human beings; we are fully implicated in it by what we are at this precise moment. We are fully and presently implicated because formulating a question about consciousness is an act of consciousness; understanding a question about consciousness is an act of consciousness; figuring out how we could answer a question about consciousness is yet another act of consciousness. In short, questions about consciousness are radically self-referential. (Bitbol, 2011, p. 1)

Now, the last sentence, about the self-referentiality of consciousness, clearly locates the author, Michel Bitbol, in one of the two major camps in which the field of the study of consciousness is divided. He is in the camp of *consciousness as primary* (Bitbol, 2011).

In fact, these two major camps are debating mostly about the origin of consciousness, in particular whether this is due to the brain, or not. Those who exclude the primary cause of the brain (mostly philosophers) are in the camp of consciousness as primary (e.g., Michel Bitbol, Evan Thompson, Mauro Bergonzi, Rupert Spira); if instead consciousness is considered dependent of the brain, or, as it is said, secondary to the brain, we are in the camp with Christof Koch, Giulio Tononi, Antonio Damasio, Daniel Dennet, and partly Aldous Huxley—mostly neuroscientists.

It is not possible to review them all here. However, let me say only that Koch (who also worked with Francis Crick; see Crick & Koch, 1990) was holding that the mechanistic basis of consciousness is a scientifically tractable problem using the tools of modern neurobiology; he also worked with Giulio Tononi, known for his *Integrated Information Theory* (IIT) of consciousness (see Tononi, 1999). They posit that what it is like subjectively and objectively is equivalent in terms of causal properties. Antonio Damasio (1999) has written a lot about consciousness, and, according to him, basically any subjective experience (an emotion) can activate a neural pattern in the brain, which develops mental images that remain in the awareness of the individual. Damasio also distinguishes between various levels of consciousness. The philosopher Daniel Dennet's book, with the ambitious title *Consciousness Explained* (1991) gave an account of consciousness in terms of various calculations occurring in the brain—the so-called multiple draft theory of consciousness. Furthermore, Aldous Huxley wrote a remarkable book, *The Doors of Perception* (2009), in which he explores the altered states of consciousness. There would be so many other interesting, fascinating authors to mention and discuss, I can only apologize for not having the space and time to do so in this paper.

Let me summarize by saying that all these last authors see, with different mechanisms and theories, the brain as the main seat of consciousness. This is quite different from the authors mentioned earlier, for whom consciousness is primary, a view which does not require a thinking brain to account for the subjective experience.

A related question is whether each of us has his/her own consciousness, or, instead, whether there is one consciousness for all individuals.

Actually, in this regard, the more general question is whether and to what extent consciousness is linked to life. It may appear strange to outsiders, but this link is not always taken for granted, also among the authors belonging to the camp of consciousness as primary.

It can be posed that, if consciousness is a personal experience, I only have experience of my own consciousness; or put differently, my personal consciousness is the only experience I have. On a personal note, this leads to postulating the simultaneous arising of life and consciousness (Luisi, 2017, 2020), meaning that each time an organism is born, consciousness is simultaneously born. This is true for a human as well as a lizard. Of course, the lizard is not aware of having consciousness, in other words, it has no self-reflexive consciousness. The idea of this simultaneous arising of life and consciousness, however, does not find much sympathy even within the camp that considers consciousness primary.

As I noted, all of this is only a fragment of what the literature on consciousness has to say, but I think it is enough to provide a framework for locating Maturana's ideas. To which camp does he belong, and what are his contributions?

Enter Maturana

As mentioned earlier, Maturana has written a few papers about consciousness, however he generally does not refer to the authors cited above, and this makes comparison difficult. What I will do is to find passages in his writing, that orient to the field in general. Let us begin saying that Maturana, as a neurobiologist, likes to emphasize the centrality of the nervous system, as he did for example in all of his older accounts on cognition (Maturana & Varela, 1980). And later on, he puts the nervous system also at the basis of mental phenomena which for him also include consciousness:

The nervous system is a structure-determined composite entity (SDS), a system such that all that happens in it and to it is determined in its structure. Nothing external to a structure determined system (SDS) can specify what happens in it. The structural changes that an SDS undergoes arise either as a result of its internal dynamics, or triggered through the encounter of the properties of the elements that compose it with the elements that compose the medium. The external elements that impinge upon an SDS do not determine the structural changes that arise in it, they only trigger them (Maturana, Mpdozis, & Letelier, 1995, p. 17)

Those are familiar terms for those who know his writings. However, how he writes about consciousness is less familiar:

> I claim that that which we want to connote when we say that a person is self-conscious does not occur as an operation of the nervous system. As all experiences do, the experience of being self-conscious occurs when the silent body sensoriality of an operation in self-distinction becomes distinguished as an element of inter-objectivity in the flow of a conversation of self-distinction. (Maturana 2005, p. 69)

Here one can already note some discrepancies with the mainstream literature. To make this clearer, we can go back to the 1995 paper written with Jorge Mpodozis and Juan Carlos Lettelier, whose abstract reads:

> To understand the biological and neurophysiological processes that give rise to human mental phenomena it is necessary to consider them as behavioral relational phenomena. In particular, we propose that: a) these phenomena take place in the relational manner of living that human language constitutes, and b) that they arise as recursive operations in such behavioral domain. Accordingly, we maintain that these phenomena do not take place in the brain, nor are they the result of a unique operation of the human brain, bat arise with the participation of the brain as it takes part in the generation of the behavioral relational dynamics that constitute language. (Maturana, Mpdozis, & Letelier, 1995, p. 15)

We see here again that Maturana is on the side of the neurobiologists, but there is a profound discrepancy with that literature as he makes human behavior and relations primary.

Let us consider again his (2005) article, "The Origin and Conservation of Self-Consciousness." This is a useful article, in the sense that you can find Humberto's whole thesis: In order to talk about self-consciousness, he has to first lay down his notion of self, time, space, the medium, structural determinism and structural coupling, and even the basic systemic laws of science, including evolution, lineage, and of course languaging, with social interactions. All that, which is the universe of Maturana, is presented in a well-written, self-consistent essay. This is important as it provides an important clue: In order to talk about consciousness, for Maturana one needs to first have a clear view of the epistemological grounds for one's claims. The other important point in this basic article, is that he talks about self-consciousness, not consciousness. This is not a trivial distinction. As I said, there is no reference in this paper to other authors who deal with consciousness and self-consciousness. This is not very surprising, but it does make it more difficult to make comparisons. But if we consider, for example, that most authors accept that the act of consciousness is an act of personal experience, Maturana's position is very different:

> Furthermore, the idea that an experience can be subjective implies the notion that an experience could be objective. From what I have said already must be apparent that I think that that which we connote as we speak of an experience is neither objective nor subjective. (Maturana, 2005, p. 74)

And finally, still in his 2005 paper, he writes something clearly outside the classic guidelines of literature on consciousness: namely, that the subjective experience may happen only when you are already in full awareness:

> I claim that that which we call experience in our daily living occurs when we distinguish what happens or is happening to us as we operate in self-consciousness: that which we call experience in our daily living occurs as our awareness of what is happening to us as a happening of our living in the moment in which live it. Without such awareness of what we are living as we live it, there is no experience. (Maturana, 2005, p. 74)

And there is also the following:

> As an operation in languaging self-consciousness occurs in the relational domain of the organism in which languaging takes place as a flow in consensual co-ordinations ... a relational dynamics of self-distinction that happens as an inter-objective entity in the present of its coexistence with other organisms. (Maturana, 2005, p. 69)

In other words, if I understand this passage correctly, first there is languaging, which for me means thinking and socializing, then the personal experience of consciousness or self-consciousness. And in later papers (Maturana, 2006, 2008), he emphasizes the same concepts. From the abstract of his 2008 paper for example:

> My reflections will be first, about how the brain operates in the generation of the adequate behavior of an organism in a changing medium, and second, about how self-consciousness appears in the course of the history of humanness" (p. 18).

Of course, one cannot talk about Maturana without mentioning Francisco Varela, and, concerning consciousness, it is easy to predict that Francisco, as a pioneer in the theory of embodiment, cannot conceive consciousness as something detached from reality, existing as an isolated essence. In fact, citing his ideas, as described by a few of his collaborators (Rudlauf, Lutz, Cosmelli, Lachaux, & Le Van Quyen, 2003), Francisco Varela's neuro-phenomenology provides an original approach to the hard problem of consciousness by having subjectivity be radically intertwined with its biological and physical roots. Citing from their abstract, "it must be understood within the framework of his theory of a concrete, embodied dynamics, grounded in his general theory of autonomous systems" (Rudlauf et al., 2003, p. 27).

And to this regard, it is also interesting, and intriguing, that Michel Bitbol (2021, p. 105) writes the following about Varela's neurophenomenology: "Varela's neurophenomenology was conceived from the outset as a criticism and dissolution of the 'hard problem' of the physical origin of consciousness. Indeed, the standard (physicalist) formulation of this problem is what generates it, and turns it into a fake mystery."

Without getting into the details of such argumentation, we can say that Maturana and Varela, basically, although they did not work together specifically on this issue, agreed on general concepts about consciousness in the sense that pure personal subjectivity is not the basic source of consciousness. Varela's name would also bring us to the notion of cognition, on which they did work together (Maturana & Varela, 1980). The notion of cognition, based originally on an analysis of the nervous system that reveals it to be a closed network of change in relation to the activity of the internal components (Maturana & Varela, 1980, 1998), is quite different from consciousness, and it was pragmatically used within the theory of autopoiesis to indicate the doing of the organism in order to maintain autopoiesis. It is not the place here to dwell further about this particular aspect of the work.

Maturana has published many papers that mention consciousness. However, I believe that the few citations I have given are enough to obtain a clear view of his position.

By Way of Conclusion

My original idea for this article, was to discuss Maturana's ideas about consciousness with respect to the main authors on the theory of consciousness. However, this idea did not really work as Maturana did not refer explicitly to people such as Michel Bitbol, Evan Thompson, Giulio Tononi, or Antonio Damasio. You may wonder why he did not, given the fact that he was indeed interested in the question of consciousness. My answer is, that he didn't care to answer them, simply because his view was based on a quite different world. His view, more or less, is that consciousness becomes the subject for discussion only after humans have formed a society where they can talk with each other, using language. On that particular point (Maturana, 2005) in fact, says: "The greatest consequence of the arising of self-consciousness and self-awareness in the constitution of humanness, is that to the extent that we human beings are self-conscious beings we are aware of what we do" (p. 72). On the other hand: "Human existence occurs in the relational and operational space of living, not in an abstract domain of consciousness, values, or intentions" (p. 73).

And all this, finally, goes back to Maturana as we know him:

> My starting point in my reflections about how I do what I do as I operate as an observer observing my making distinctions, is myself, a human living being that does not pre-exist to his or her self-distinction. What is fundamental and particular in this starting point, is that it does not entail any ontological assumption, and the observer must show how he or she operates as a self-conscious living human being with the operational coherences of his or her operation as a self-conscious living human being ... I have written elsewhere "Everything said is said by an observer to another observer that could be him or herself." This is a constitutive condition in our operation as human beings. (Maturana, 2005, p. 57)

And let me add his perspective on human existence and the place of consciousness in it:

> Human existence occurs in the relational and operational space of living, not in an abstract domain of consciousness, values, or intentions. All that we live occurs in us human beings through the recursive historical transformation of our body hoods and of the configurations of sensory-effector correlations that we generate, and we live what we live as a world that arises moment after moment in a historical operational flow as an expansion of the transforming body hood that we continuously become through our coexistence in a changing languaging community. (Maturana, 2005, p. 73).

I will end with the last book by Humberto Maturana, written with Ximena Yanez (2017), a book in which the word consciousness is very common, and where, talking about man and humanity, they introduce the concept of Homo sapiens amans amans. One of the last chapters ends with an important message for all of us:

> In the cultural period in which we live, in the consciousness space in which humanity lives in this globalized world ... every cultural transformation begins as an individual transformation. And ... the future of the living humanity is not a responsibility of our children, but it is of the young people of today and of the less young ones who are living now with them. (Maturana & Yanez, 2017, p. 472; my translation)

Acknowledgments

The author thanks with gratitude the comments of Jay Efran and Fred Cumming

References

Bergonzi, M. (2011). *Il sorriso segreto dell'essere*. Oscar Mondadori.
Bitbol, M. (2011). On the radical self-referentiality of consciousness. In R. Penrose (Ed.), Consciousness and the universe (pp. 99–107). Cambridge, MA:The MIT Press. (This text —presented in some conferences—picks out the philosophical nucleus of M. Bitbol & P.-L. Luisi, [2011]. Science and the self-referentiality of consciousness. *Journal of Cosmology, 14*, 4728–4743. It contains a short outline of a more intricate argument to be found in chapter 1 of M. Bitbol [2014]. La conscience a-t-elle une origine? Flammarion.)
Bitbol, M. (2021) The tangled dialectic of body and consciousness: A metaphysical counterpart of radical neurophenomenology. *Constructivist Foundations 16*(2), 141–151.
Chalmers, D. (1995). *Journal of Consciousness Studies, 2*, 200–219.
Crick, F. (1994). The astonishing hypothesis. London: Touchstone Books.
Crick, F., & Koch, C. (1990). *Towards a neurobiological theory of consciousness. Seminars in Neuroscience, 2*, 263–275.
Crick, F., & Koch, C. (1998). Consciousness and neuroscience. *Cerebral Cortex, 8*(2), 97–107. https://doi.org/10.1093/cercor/8.2.97
Damasio, A. (1999). The feeling of what happens. New York: Harcourt Brace & Co.
Dennett, D. (1991). *Consciousness explained*. Little, Brown and Co.
Humphrey, N. (2009). *Seeing red, a study in consciousness*. Belknap Press.
Huxley, A. (2009). *The doors of perception*. Harper Collins.
Luisi, P. L. (2017, March 23). *Wall Street International* (p. 1).
Luisi, P. L. (2020). *Essays on life sciences, life and consciousness: A simultaneous origin?* Cambridge Scholars Publishing.
Maturana, H. R. (1985). Comment by Humberto R. Maturana: The mind is not in the head. *Journal of Social and Biological Structures, 8*(4), 308–311. Available at http://cepa.info/575 (Discussion of Aboitiz F. D. (1985). A critique of the modern concept of localization. *Journal of Social and Biological Structures, 8*(4), 307–308).
Maturana, H. R. (1987). The biological foundations of self consciousness and of the physical domain of existence. In E. R. Caianiello (Ed.), *Physics of cognitive processes* (pp. 324–379). Singapore: World Scientific.
Maturana, H. R. (1990). *Biology of cognition and epistemology*. Temuco, Chile: Ed Universidad de la Frontera.

Maturana, H. R. & Varela, F. (1980). *Autopoiesis and cognition: The realization of the living*. Boston: D. Riedel.
Maturana H., & Varela, F. (1998). *The tree of knowledge*. Boston: Shambala.
Maturana, H. R., Mpodozis, J., & Letelier, J. C. (1995). Brain, language and the origin of human mental functions. *Biological Research, 28*, 15–26.
Maturana, H. (2005). The origin and conservation of self-consciousness: Reflections on four questions by Heinz von Foerster. *Kybernetes, 34*, 54–88.
Maturana H. R. (2006) Self-consciousness: How? When? Where? *Constructivist Foundations, 1*(3), 91–102.
Maturana, H. R.,& Yanez, X. D. (2017). *El Arbol del vivir*. Santiago: MVP-Editors.
Maturana, H. R. (2008), Anticipation and self-consciousness. Are these functions of the brain? *Constructivist Foundations, 4*(1), 18–20.
Rudlauf, D., Lutz, A., Cosmelli, D., Lachaux, J.-P., Le Van Quyen, M. (2003). From autopoiesis to neurophenomenology: Francisco Varela's exploration of the biophysics of being. *Biological Research, 36*, 27–65.
Thompson, W. I. (2006). *Coming into being*. New York: S. Martin Press.
Tononi, G. (2004). An information integration theory of consciousness. *BMC Neuroscience, 5*, 42.

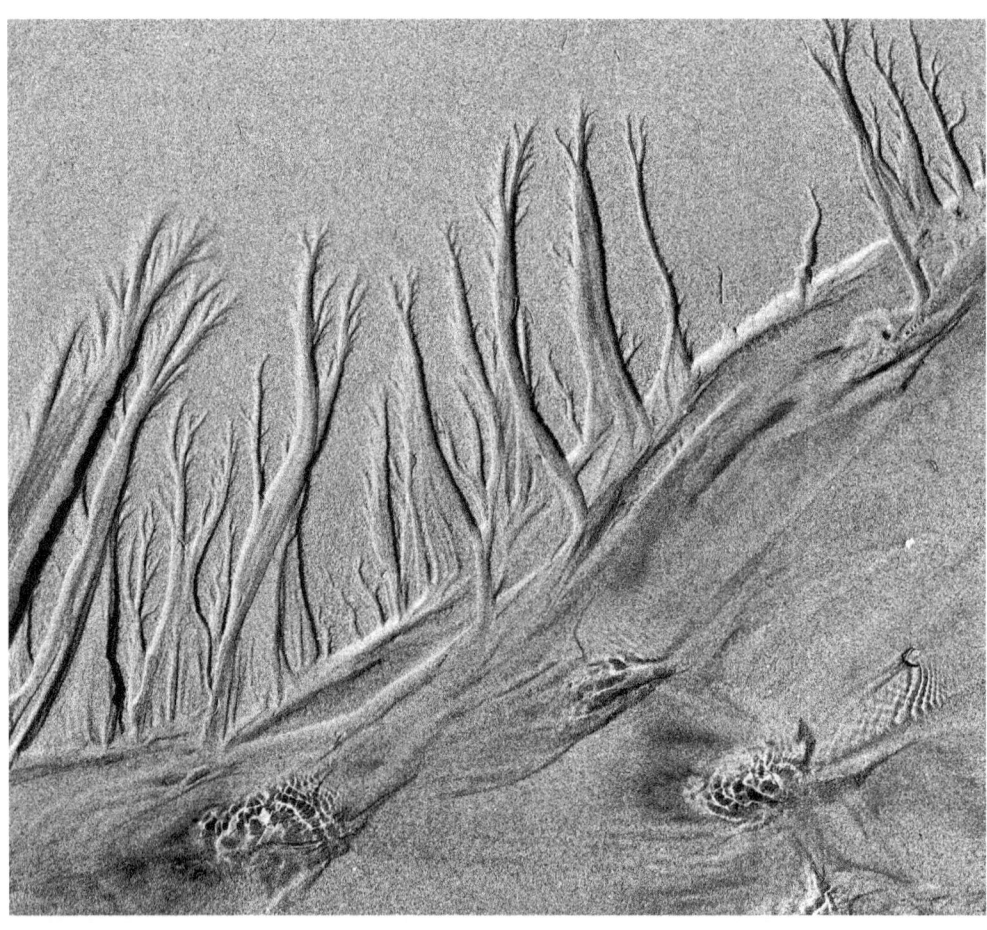

Bunnell, P. (2019). *Sand Forest*. Photograph.

Humberto Maturana on Time:
Zero-Time Cybernetics

Jude Lombardi[1] and Larry Richards[2]

The American Society for Cybernetics (ASC) awarded Humberto Maturana the Norbert Wiener Medal in 2008. Since he could not attend in person, Maturana sent a letter to the ASC in which he proposed a zero-time cybernetics. After offering some personal reflections on our interactions with Humberto, the authors of this paper engage in an exchange of ideas on time, zero-time and zero-time cybernetics. Our hope is that Humberto Maturana's letter and our thoughts might inspire more discussion and conversation on these topics within the cybernetics community.
Keywords: Humberto Maturana, zero time, structural dynamics, languaging, conversation, zero-time cybernetics, presence-oriented consciousness, biological/cultural matrix, biology of love

Introduction

Both authors of this paper (Jude Lombardi and Larry Richards) have a positive history with Humberto Maturana. While we each had a different relationship with him, we came to admire him as a thinking, caring person who made huge contributions to biology and cybernetics. We wish to honor him and his contributions with some personal reflections and an exchange of thoughts and questions on one of his papers, in particular, a letter that he sent to the American Society for Cybernetics in 2008 upon receipt of the ASC Norbert Wiener Medal, awarded for lifetime contributions to the field of cybernetics. In his letter, Humberto offered some thoughts on the concept of time, specifically invoking the idea of zero-time (or no-time). We think this topic has obvious and profound implications for philosophy. However, he offers it in the context of science, especially biology and what he has called the biological-cultural matrix, where we think it represents new and hitherto unexplored territory. We will let the letter speak for itself in that regard. We decided to focus our back-and-forth exchange on implications for cybernetics in general, drawing on his earlier paper "The Nature of Time" (Maturana, 1995), to help in sorting out the many issues and nuances of Humberto's insistence on *zero-time* and his proposal for a *zero-time cybernetics*. We make no claim to understanding all the issues and nuances of Humberto's intent; rather, we devote our attention to identifying some questions and responses in the hope of engaging others in this exploration of new directions in science and life.

Humberto Maturana's (2008) Letter to the American Society for Cybernetics

> First of all, I wish to thank you for the distinction that you wish to bestow on me. It is several years that we do not see each other, but your friendship is dear to me. Now as an act of appreciation and

---

1. Videographer, social worker, artivist. Baltimore, Maryland, USA. Email: jlombardi@jlombardi.net.
2. Dialectician, organization designer, conversationalist. Portland, Maine, USA. Email: laudrich@iue.edu.

respect to you I wish to say a few words in relation to what I would have liked to present to you if I had been able to come. And I would like to speak about a notion that I and my colleague Ximena Dávila Yáñez have been developing since we created together the Matriztic Institute in Santiago the year 2000, and that is called "Biologico-Cultural Matrix of Human Existence". I shall not develop this notion in full in this short note, but I shall speak of it referring to its conceptual-operational roots.

All the work that Ximena and I are now doing entails the understanding of the biology of cognition and the biology of love, and I shall synthesize this understanding with the presentation of three systemic laws:

Systemic law 1: "Everything said is said by an observer to another observer that could be him or herself." The observer is a human being or some other being operating as a reflective human being in language.

Systemic law 2: "Whenever in a collection of elements some configuration of relations begin to be conserved, a space is opened for everything else to change around the configuration of relations being conserved". This is the spontaneous manner of arising composite entities.

Systemic law 3: "The result of a process does not ever participate in its genesis."

Systemic laws are not ontological assumptions, nor are they definitions; they are abstractions that the observer makes of the configurations of operational coherences that he or she distinguishes in the different operational-experiential domains [in] which he or she operates in his or her living. So, they apply everywhere in the cosmos that the observer brings about with his or her living. Indeed, in this respect Systemic Laws are not different from any other law of nature since they arise in the same way and apply in the same cosmos.

At the end of the year 1999, Ximena (who worked then as a family consultant) approached me saying: "Prof., I have made a discovery; I have discovered that all the pain and suffering for which one asks relational help, is of cultural origin in this patriarchal culture in which we live." And then she added: "Moreover, as the consulting person tells me of her or his pain, she or he unconsciously reveals to me the moment in the cultural relational matrix of her or his living where the pain and suffering that she or he is now living originated; and in the same process she or he unconsciously reveals me also the path out of such pain and suffering in the cultural relational matrix that she or he is living now." And in the flow of our conversations along the months she showed me that the pain arose in a moment of denial of love in the past but was not of the past because it existed in being continuously conserved in the present. This was a remarkable assertion that I did not take lightly. When Freud introduced a social view in his study of hysteria, he spoke of trauma and of repression; Ximena spoke of the negation of love in the cultural domain, and spoke of conservation of pain in the present.

We human beings exist in the present, in a continuously changing present, the past and the future do not exist as such, and they are manners of being now, in the present. The cosmos that we generate in our living occurs, exists, as a continuously changing present. The past is a way of explaining the present being lived arose in its continuous change by proposing a generative mechanism that would have given rise to it if the operational coherences of the now being lived were conserved. The future is a manner of living now in the proposition of what would happen if the operational coherences of the present being lived now are conserved in the continuously changing present being lived. Autopoiesis, living, occurs as an in a continuously changing present: living occurs in no time, in zero time. But what is time, then? Is it not time one dimension of the physical space? What are we saying when we speak or talk about regulation and control?

These questions lead me to the following reflections and conversations with my colleagues of the Institute:

When we speak of regulation and control, we intend to relate logically processes that occur in non-intersecting domains and which can only be correlated historically through the memory of their repeated but independent observation. In these circumstances, our ancestors using memory as an operational referent, invented time as an imaginary spatial dimension that would allow them to connect semantically otherwise not related events that result in the historical structural change of a system. Let me use a pressure-cooking pot as an example to illustrate what I mean. It is usually said that in a pressure-cooking pot the "rider" that is placed on an opening on top of the lid of the pot, operates as a valve that lets the steam out to regulates the temperature of the water in it. What actually occurs is that the pressure-cooking pot with water in it and under the fire, has a dynamic architecture whose structural changes result in some independent events occurring in several non-intersecting operational domains that an observer is able to correlate, and then put together in what seems to be a logically coherent causal story in the domain of cooking by saying that the rider regulates the temperature in the cooking process. The observer can do this only after inventing an imaginary spatial dimension that would allow him or her to connect through his or her memory as a single event process that he or she has lived (or has imagined to have lived) in different presents that are otherwise unrelated.

I do not say that our ancestors reflected as I am doing now when they invented time. It was not necessary for them to do so, they just languaged it in their living together as a particular coordination of doings in the flow of their recursive coordinations of coordinations of doings as they lived together as languaging beings. Moreover, in doing so they generated time in the same spontaneity as they generated each one and all the things, entities, notions concepts, ... of the worlds that they lived as they live together their languaging human living.

Time as an imaginary spatial dimension transformed the human operational-conceptual world as other imaginary notions like imaginary numbers have also done. I think that the invention and use of time as an imaginary new dimension of space in the three-dimensional present in which our non-languaging existence occurs, permitted us cultural human beings to conceptually and operationally connect not-connected processes and events occurring in independent domains by creating the operational domain of descriptions as a manner of living in the flow of events in their succession in "time". The semantic notions that the imaginary dimension of time permitted to introduce in the description of the operation of systems became an operational dynamics connecting process that occur in domains that do not intersect because they occur in different presents. The use of semantic notions seemed to facilitate the understanding of the operation of systems by treating processes that occur in non-intersecting domains as if they occurred in our daily life which is where semantic notions operate.

The beauty of using the imaginary dimension of time in the description of what we do, we think, or we see in the flow of our living, is that it permits us to propose connections in the present that we are living with whatever we imagine that we are living or that have lived, regardless of whether they occurred in the same or in non-intersecting operational domains. The use of the imaginary spatial dimension time permits us to reflect on what we do and feel as if we were, without knowing it, in a shadow theater, relating things that we do not see that cannot be related because they occur in different non-intersecting operational domains. To do this, though has had an undesired additional consequence, namely, it has obscured our understanding of the architectural dynamics, which as a flow of structural change in the continuous changing present of existence, gives its historical unity to a system that in the flow of its own existence occurs without past and future as a continuous now. The most usual and difficult trap that comes from such blindness on the structural dynamics of systems is resorting to reductionism in an attempt to escape the realization that past and future are cultural manners of living in the present.

We human beings cannot live as cultural human beings without time, but we live our biological continuous changing present without time, that is in zero time, as all living beings do even when we use the notion of time as a physical dimension, forgetting that it is an imaginary explanatory spatial dimension invented to bound the beginning and the end of systems in their operation as discrete entities. The cosmos that we bring about as cultural human beings exists in time, but all the process that we describe as occurring in by themselves in it occur in zero time. Organisms operate as discrete singular entities, or totalities, in a continuously changing present, and they exist as totalities bounded by borders generated through their own operation, without end or beginning because they occur outside our description in zero time. It is we cultural humans that exist in time who want a temporal closure for organisms and wish to see them with beginning and end. But to do so we have to use time leaving out for a while semantic explanatory notions or semantic connections and look at the organisms as dynamic architectures and see them as a continuously transforming structural dynamics in the present. If we manage to do so in our imagination, we shall indeed see organisms as self-bounding four dimensions autopoietic totalities with the form of a sausage that is beginning at its conception, and that is ending at its death.

This can be illustrated with a drawing of a vertical sausage of processes to be looked through a slit that moves from its beginning to its end (or birth and death in an organism). [See illustration and instructions below – Ed.] If we do this, we will see at every moment as we look through the sliding slit a changing present disconnected from all other moments of changing present that we have already seen or that we think that we will see later. If all that we see in our living is a slice of our continuously changing presents, the imaginary spatial dimension of time offers us two basic possibilities to interconnect the separated presents of our living. One is to interconnect them with a thread of semantic notions such as control, information, purpose, or regulation, that are semantic notions that we easily use in our daily life in occasions where memory helps us not to confuse the description of a process with the process itself. The other basic possibility is to orient ourselves to see the changing dynamic architecture that the organism, with everything else in the cosmos or worlds that we bring about in our cultural living is. I consider that this second possibility is the one that will lead us to the understanding of our existence as cultural human beings because it will lead us to see how not-intersecting relational-operational domains arise continuously with what we do, as the recursive dynamics of the spontaneous generation of intrinsic novelty in the cosmos and worlds that we bring forth with our cultural living. Indeed, non-intersecting domains arise as we do distinctions, reflections, recursive operations, and we mostly do not see this because we usually treat what we do as if they were happening as part of the totality of our doing. As this happens to us, we usually correlate the independent processes in the not-intersecting domains without being fully aware that they do not intersect, and we propose logical relations between them which do not apply under the form of ad-hoc semantic relations. When we attend at the dynamic architecture of what we are doing we do not confuse domains and do not treat correlations as if they were logical relations resulting from their occurring in the same operational-relational domain. And finally, if we attend to the dynamic architecture of systems, we will also realize that as they exist in a different domain than the domain of existence of their components, they affect each other through the operation of their components if these exist in the same domain. Much of the confusions and misunderstandings in relation to the use of notions such as language, consciousness, or mind arise from the use of semantic notions that obscure the vision of the changing dynamic architecture of the systems and the worlds that we generate in our human cultural existence.

We cultural human beings use in our daily living many imaginary explanatory notions, but whatever these may be, they operate through the realization of our living as autopoietic organisms in a space with four dimensions, one of which is imaginary time. But it does not matter that time should be an imaginary dimension, autopoiesis, living, occurs as a changing dynamic architecture, and as a result of that, all that we human beings do, regardless of whether we do what do as biological or cultural beings, occurs in one single domain, in the flow of our realization as autopoietic beings. We, cultural human beings, as all living beings, live as valid all that we live, and this is how we live in the same manner both imaginary and not imaginary spatial dimensions. That is, as autopoiesis occurs in a domain of three spatial dimensions plus one imaginary that is time, and all that occurs in and with living systems, and particularly all that occurs in and with us occur in the realization of our living, all other imaginary relations that we create in our cultural living become valid for our living in zero time that we are living, whatever their historical consequences in other zero time in the flow of the cosmos that we bring forth with our living.

My colleagues and I in the Institute are using the understanding of the changing dynamic architecture in our living to understand the actual operation of systems without introducing semantic notions to explain what happens in them and with them. And in particular to understand how does the past that we generate as an explanatory notion in an imaginary dimension, operates in the zero time of the present conserving cultural pain and suffering.

In the course of our living we have just published in Spanish a book called "Habitar Humano, en seis ensayos de Biología-Cultural." The title in English will be: "Human Habitats, in six essays of Cultural-Biology".

Many thanks,

Humberto Maturana Romesín (May 2008)[3]

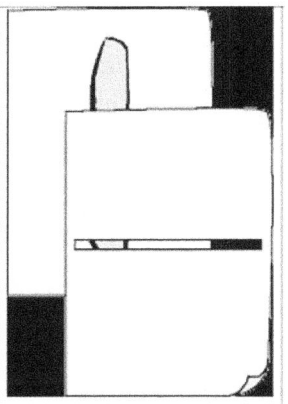

The 'sausage and slit' illustration can be demonstrated using two pieces of paper.

On one is a picture of a vertical 'sausage'- shaped figure representing the organism's totality in four dimensions (including time). The 'beginning' is toward the bottom and the 'end' is toward the top.

The second sheet of paper has a narrow horizontal slit through it.

Moving the slit sheet over the figure sheet illustrates (a) the limiting scope of the instantaneous present and (b) how the ascribed 'future' and 'past' ('sausage' portions visible above / below the slit sheet) lie outside this scope.

Sausage and Slit Illustration (Maturana, 2008, p. 5)

Personal Reflections

Larry Richards: I first saw Humberto give a presentation at a Gordon Research Conference in New Hampton, New Hampshire, in 1984. I would see him many times again at conferences of the American Society for Cybernetics (ASC) and other organizations and had a chance to interact with him on some of those occasions. During the time I served as President of the ASC (1986-1988), he presented at five conferences that I attended. I struggled with the ideas he was presenting, at one time telling my ASC colleague, Rodney Donaldson, that I did not see the significance of these ideas to anything in which I was interested. Rodney, to his credit and with my

---

3. https://asc-cybernetics.org/2008/HM-08WienerComments.pdf

gratitude, would not give up on me. We had many conversations, and I came to have a great appreciation of Humberto's contributions, not just to biology, but to the human enterprise in general and our lives in it.

One encounter stands out for me. At the ASC conference in Victoria, British Columbia, in 1988, Humberto made three comments that stuck with me and still today gnaw away at my brain.

- A couple of us happened on Humberto and Gordon Pask having dinner and asked if we could join them. At some point, the topic of language came up. I suggested that the phenomenon that Humberto called languaging was similar to the phenomenon that Gordon called conversation, to which they both immediately responded "No." This was an important moment for me, as it helped clarify both concepts. Languaging is about coordination, specifically, the consensual coordination of the consensual coordination of action. Conversation is not about coordination, even though it happens in language and some coordination might be useful or necessary. Conversation arises from frictions (that is, asynchronicities) among the participants and proceeds as an intimate interaction as they work to resolve their differences. Conversation is the converse of control.
- During his conference presentation, Humberto was asked what he would recommend to mitigate the cruelty and violence so prevalent in human societies today. His response was something to the effect that we should let biology work. His argument was based in part on what he calls the fundamental emotion of mammals in general and humans in particular, namely love. The cruelty and violence we witness in the world is a consequence of the denial of love that characterizes much of the day-to-day systems in which we work and play. Does this imply, I thought to myself, that we shouldn't do anything about poverty, oppression, environmental degradation, and so forth, but rather just let it work itself out? Not necessarily, but perhaps the setting of goals and seeking of solutions contribute to the very phenomena we are trying to avoid.
- In private, I asked Humberto about how he addresses in his formulations the concepts of time and history. Certainly, biology and science in general depend on these concepts, yet they are abstractions that we take for granted in our everyday lives. I do not remember his exact response, but I seem to recall that he indicated that time was a human invention and that history was a necessary concept for explaining the world. However, all we know and all we do is right now, in an everchanging present. The past explains what is now; the future explains what is not now. It was this interaction that led me to the idea of the cybernetician as a craftsperson in and with time (Richards, 2020).

Jude Lombardi: Conversation as the converse of control entails letting go and letting biology work.

Your statement "Conversation is the converse of control," reminds me of Humberto's comment in his 2008 letter regarding a relationship between time, control and regulation. All of which, he claims, are external notions that require the invention of an imaginary spatial dimension. He further suggests that our ancestors did not likely do this intentionally, but that in the flow of their living together as languaging beings, in their particular flow of their recursive coordinations of coordinations, such notions emerged out of necessity. Our human need for order is met by the necessity for time?

I too realize time is a necessary human invention and that models of consciousness that are purpose oriented and goal directed are necessary, but also limit our doing in certain domains, particularly the domain of dynamics where biology is a happening. Biology, in this context, is not about the study of life but a study in living. Human living always takes place in the relational space of humanness where culture and biology intersect. We humans exist in zero-time in the continuous flow of our changing conversational presence. So as Humberto said: "Your present changes your past changes. Your present changes your future changes" (Maturana, 1992a, quotation starts at 1:50 min). How hopeful.

Larry: What is your history with Humberto?

Jude: I first met Humberto Maturana along with Herbert Brün (both served on my doctorial committee) at my first American Society for Cybernetics meeting in 1992. The conference took place on a small, isolated island in Seabeck, Washington state. It was organized by then ASC president Rodney Donaldson and entitled "Language, Emotion, the Social, and the Ethical, An In-depth Exploration of the Cybernetics of Herbert Brün and Humberto Maturana." I had no idea what I was getting into and would soon find out.

I knew intuitively, they were saying things radical and unique that I needed and wanted to hear and yet I could not comprehend. It was also during the 1992 conference that I emerged as a video ethnographer. I videoed much of the conference which allowed me to view the lectures over and over again, learning more about these radical ideas with each view.

Recursion: Always Back on Self Yet Never Quite the Same.
At the 1992 ASC conference, both Brün and Maturana offered me what I was searching for, healthy models for understanding and explaining human thinking and doing. Back then I was a social worker, working with severely "emotionally disturbed" children. My particular specialty was the de-escalation of behavior. I had first heard the term *cybernetics* while attending a training by Allen King Cooper who was hired by the state of Maryland to educate childcare professionals in how to talk kids down without violence. Cooper's ideas led me to create *behavioral cybernetics*, which eventually led me to the ASC in 1992.

During the 1992 conference, I clearly remember Humberto drawing his closed circle, representing the autopoietic system, in a relational medium, with an observer. (Maturana, 1992b)

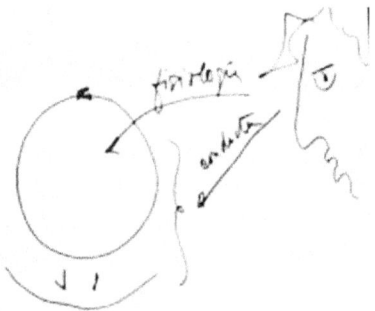

Image from video of Maturana's presentation "Relativity" (Maturana, 2012; https://www.youtube.com/watch?v=5WgpP0MfHXc; image appears 18:44 mins)

I thought his explanation was more natural than any of the others I had ever experienced. During his presentation he pointed at a variety of dynamic/relational explanations for living. I will focus here on three fundamentals that orient my thinking 30 years later. They are:

I. His cybernetic starting point regarding humans as observers observing.
II. His biological explanation of living beings as fundamentally emotional beings.
III. His claim that the biology of love is the fundamental emotion of all living beings.

I. His cybernetic starting point. Observers live immersed in language. It is our living immersed in language that generates us as observing beings. This makes humans unique and also allows us to articulate what makes us similar to other living systems. "We human beings happen in language, and we happen in language as the kind of living systems that we are. We have no way of referring to ourselves, or to anything else, outside it. Even to refer to ourselves as non-languaging entities we must be in language. Indeed, the operation of reference exists only in language, and to be outside language is for us as observers, nonsensical. For these reasons it is essential for understanding the observer as a human being, to explain language as a biological phenomenon." (Maturana, 1992b, p. 22)

II. His biological starting point. All living creatures are molecular autopoietic organisms immersed in a medium at every moment of their living. It is in this biological cultural matrix of humanness where an observer observes body dispositions that emotioning emerges.

> Emotioning: any and all body dispositions observed in the relational space of humanness.

It is emotioning that constitutes living. No emotioning, no living. Emotioning is a dynamic all living systems share. We humans live in a dynamic emotional/relational space of humanness. Maturana spoke often about emotions, moods and emotioning:

> The Western culture to which we modern scientists belong, depreciates emotions, or, at least considers them a source of arbitrary actions that are unreliable because they do not arise from reason. This attitude blinds us to the participation of our emotions in all that we do as the background of bodyhood that makes possible all our actions and specifies the domains in which they take place. And this blindness, I claim, limits us in our understanding of social phenomena. Let us reflect upon this matter: (i) All animals have different domains of internal operational coherences that constitute dynamic body postures through which their actions and interactions in their respective domains of existence take place. (ii) The observer distinguishes different emotions and moods through the distinction of the different domains of actions in which the observed organisms move. Thus, all animal behavior takes place in a domain of actions supported and specified at any moment by some emotion or mood. Indeed, all animal life takes place under a continuous flow of emotions and moods (emotioning) that changes the domains of actions in which the organisms move and operate, in a manner that is contingent to the course of their interactions. We human beings are not an exception to this. Moreover, in us human beings emotioning is mostly consensual, and follows a course braided with languaging in our history of interactions with other human beings. (iii) The observer distinguishes different emotions and moods through the distinction of the different domains of actions in which the observed organisms move. (Maturana, 1992b, p. 27)

I spoke with Humberto in 1994, when he clarified for me his use of the words love and aggression.

> Love: The domain of those behaviors through which the other arises as a legitimate other in coexistence with oneself.

> Aggression: The domain of those behaviors through which others are negated as legitimate others in coexistence with oneself.

III. He goes on to claim that when observing the relational space where human living is happening, not only is emotioning a fundamental feature but that love is the fundamental emotion of all living creatures—in the domains in which they exist.

> The biology of love: the domain of behaviors in which the other arises as a legitimate other in coexistence with oneself.

At the 1992 conference, Humberto told the story of a spider and a human and their relationship as an example of the biology of love across species (Lombardi & Maturana, 2021).

I consider these three fundamentals extremely important when exploring ways to address our asynchronicities without violence. In this context, it is important to articulate a distinction between Maturana's definition of *conversation* and your definition, Larry, for *conversation*, since I find them both useful in certain situations. By the way, I choose to use the term languaging to describe the braiding of emotioning with language that emerges through the domain of behaviors in the relational space of humanness.

Maturana describes conversation as a human interaction that entails the braiding of languaging and emotioning in recursive coordinations of coordinations of consensual behaviors. He also claims that all human activities take place as networks of conversations and that closed networks of conversation generate culture (Maturana, personal communication, 1996).

I also cling to your interpretation of the term conversation. What I call *deep conversation* which emerges when observing asynchronicity that evolves into synchronicity. As you have said, three categories for the word conversation are useful: a conversation we have with ourselves, a conversation we have with another, and a conversation we have with society. In each case, the *we* is a role or perspective temporarily taken (Gordon Pask's p-individual). In each case, the dynamic is similar—a back-and-forth interaction in a language starting with an asynchronicity moving toward synchronicity. (Richards, 2010). For me, this is when the concepts of zero time or no time become invaluable.

Larry: I tend to use the word *conversation* in the way you might use *deep conversation*. Sometimes, I will speak of the cybernetic version of conversation to distinguish it from common usage: what I would call polite cocktail conversation or chit-chat (which can, of course, also be useful). The cybernetic version of conversation requires a conflict, disagreement, friction, inconsistency, that is, being on a different page or out of synch with the other. However, it also requires a preference on the part of the participants for recurrent interaction, that is, a desire to continue the conversation and pursue some insight into mutual differences.

And yes, I claim that conversation, as a dynamic, happens in zero-time, not in the past or in the future, but right now, everchanging. When a conversation gets recorded as a dialogue (as in a written transcript), it is no longer a conversation. What gets recorded is the content; what does not get recorded is the dynamics. While a writer can give clues about the dynamics, and a reader can imagine a dynamics, the dynamics of the original conversation is unique to that moment. Furthermore, as an avenue to a form of participation that requires no ability to cause things to happen, this perspective is essential. Every moment I am in a conversation I make a difference in the world.

Jude: Yes, every moment in conversation I makes a difference. One possible triadic model for maintaining one's cognitive equilibrium: adaptation, assimilation, accommodation (Glasersfeld, 2003). Maybe deep conversations invite participants to accommodate rather than assimilate.

Assimilating is when what I hear fits with my established epistemological and structural dynamics at that moment. Whereas accommodating requires a deep shift in my current epistemological and structural stance and the necessity for generating a new scheme. So maybe deep conversations provoke deep epistemological shifts in one's knowing and doing?

I tend to think of conversation as a description, an abstraction, of our doing. Whereas zero-time, always in a presence, is indescribable. Our biology is such that we do the only thing we can do, always in a present. And as observing beings we operate while living immersed in languaging, which allows us to construct and explain our experiences. There are at least two phenomenal domains for observing: the domain of experience which is dynamic thus unexplainable, and the domain of explanations, using the language of logics and causal relations.

Being in zero-time is unexplainable. I wonder, if there is a possible distinction between zero-time and zero-time cybernetics?

Zero-time, a momentary phenomenon that entails an observer in their present niche. "Performance, sharing your presence" (Herbert Brün in Lombardi, 2021; quotation starts at 36:25).

Zero-time cybernetics, transdisciplinary explanations for embracing zero-time. Hence, instead of offering an epistemological stance for a consciousness that is purpose oriented and goal directed, cybernetics offers one that is also presence oriented and process directed.

Zero-Time Cybernetics

Larry: I think the last time we saw Humberto in person, Jude, was at the 2012 ASC meeting in Asilomar, California. He gave a presentation there called "Fundamental Relativity" (Maturana, 2012).

I think it was one of his colleagues at the Matriztic Institute who also gave a presentation that included a reference to zero-time, which I could not attend. Since I had missed the presentation, I decided to send Humberto an email asking about it. He replied, indicating his continuing interest in this idea of zero-time (or no-time), implying that he was focusing much of his attention on it. He attached a draft paper that was a slightly expanded version of his letter to the ASC in 2008 except he now used the term *zero-time cybernetics*.

As an example of the expansion, he had added a fourth systemic law: "Everything that happens occurs in the instant that it occurs as it occurs, not after or before" (Maturana, n.d.). Hopefully, we can exchange some thoughts and/or questions on this idea.

Jude: Zero-time cybernetics offers a model of consciousness that is present oriented and process directed in the here and now, which is all there is anyway. This reminds me of Annetta Pedretti in two ways. First, at the 1993 ASC meeting in Philadelphia, Pennsylvania, Annetta talked about the possible consequences of a temporal dimensionality when trying to describe language. "When I am weaving this thread, when I am following the thread, it becomes possible at any moment in that continuity to say the next thing. It's only a question of how I get there. It's not a question of what things are. It is a question of how I learn. It is a question of when I can say something … . As we become confident of that, we stop being these outside observers, we stop being these controllers and we start swimming in something which flows … . There is a sense in which I can trust that if you cut my thread even that can become the thread that I can pick up again and make it possible to say the thing." (Pedretti, quoted in Lombardi, 2021)

Secondly, her idea that as a society we are running out of clockwork, so we have more time for artwork. "To make things work is the work of art" (Pedretti, quoted in Seaton, 2021).

Larry: In the brief email exchange I had with Humberto following the 2012 conference, he seemed to imply that this idea of zero-time (or no-time) was what he was devoting himself to and wanted to get others involved. I was not, and still am not, sure what to do with it, except that I have had an obsession with the idea of time as far back as I can remember, which motivates in part my continual interest in cybernetics (Richards, 2016). I speak of the cybernetician as a craftsperson in and with time, implying that time is not external or given and that it can be manipulated. This is not how science treats time. I claim, as does Humberto, that time is a human invention, a concept needed in order to connect phenomenal domains that would otherwise be incompatible or opposing, and so avoid inconsistency, paradox and/or contradiction. I regard Humberto's distinction between the domain of dynamics (experience) and the domain of explanation (relations) to be a contribution, the significance of which has yet to be realized, in biology, science, policy-making, design and everyday life. So, yes, I would like to see the cybernetics community take up Humberto's proposal for a zero-time cybernetics.

Jude: Yes! A braiding of the domain of experience (dynamics) where our structural determinism is an autopoietic happening, and the domain of explanations (relations) where our languaging generates a biocultural matrix of doing.

I think zero-time cybernetics is not only important but revolutionary in that it requires letting go of notions such as control, regulation and purpose when interacting with other living (autonomous) systems. Yes, we need an external notion for time. We also need to embrace zero time as a matter of our living as autopoietic biological cultural beings doing the only thing we can do at any instant of our living. Maybe this is also something we share with other living organisms?

What do you think about the importance of the notion zero-time cybernetics?

Larry: Maturana starts with the idea that living happens in the moment, not in the past or future, but rather in an everchanging present. Hence, in no-time. He then seems to suggest that many (even most) of the concepts/explanations/descriptions we language into being depend on a concept of time (although there are significant differences across cultures). Time has become so embedded in our cultural lives, and our relation with the world, that we cannot live without it. The cybernetic ideas of goal and regulation, for example, depend on time. He suggests that, if we were to set the idea of time aside and focus on the structural dynamics of human experience, we might arrive at new insights into the human condition, while recognizing that we cannot just stop living in the world of time that we have. Perhaps, we go back and forth between the two domains, retreating to the dynamics of our being (and our conversations) when the violence of the world seems too overwhelming to address in the usual, purpose-oriented (and time-oriented) way. This seems to be related to your idea of a consciousness of presence, as opposed to the consciousness of purpose that dominates how humans tend to address problems of the world. However, if I am living in the moment, with no past and no future, there are no objects, no problems, no solutions, only dynamics. Can I even think about an everchanging present without moving to the domain of relations?

Jude: Maybe our traditional use and need for the notion of time is why wicked problems are wicked?

Larry: For me, the question is: How does the way of thinking that guides our behavior shift when we acknowledge the two domains, and especially when we acknowledge our structural dynamics in zero-time? If we take the view of an everchanging present, how can we think of change without a concept of time? Doesn't change happen over time? What if change was taken as fundamental—not object, not entity, not matter or energy or even information, but change? What world would we bring forth if the logic of the world was a logic of change and only change, in the moment? Since all currently best available knowledge is formulated with time as a given phenomenon, we perhaps have no choice but to address the problems we face with that knowledge and in the context of time. Hence the dilemma: needing to address the problems of the world in time (because that's what we know and live) while recognizing that the sources of those problems may be in how we think of ourselves and our world as dependent on a time that is external and given. What does a shift to a consciousness of presence do for us and our world? That's the question for cybernetics.

Jude: I feel like I, we, just came full circle—how cybernetic. I am beginning to see where no-time might fit for me.

I think the implications of living with awareness of the possibilities of zero-time cybernetics are magnificent. This conversation takes me back to my comment about when I was first exposed to cybernetics and my co-construction with childcare workers and teachers of behavioral cybernetics, about how to de-escalate a situation without violence. This requires that one access both their relational and dynamic cybernetic ways of thinking, being and doing in order to be an effective change agent (Lombardi, et al., 1993).

Larry: Yes, without time, there would be no problems. We would just do what we do when confronted with obstacles or threats, like all living systems do. However, we humans also live in a culture, in the context of time, and so must deal with wicked problems as a human undertaking. Jude, do you have further thoughts on the potential significance of the idea of zero-time cybernetics? I am still trying to wrap my mind around the idea and its implications.

Jude: In his 2008 letter Maturana mentions his partner Ximena's discovery that the suffering her clients were experiencing could be traced to moments in their living where the biology of love had been negated in their biocultural matrix. Furthermore, it was through the flow of their conversations that a path toward recovery emerged. I suggest this process of flowing in conversation in zero-time cybernetics while in the biology of love is fundamental. That is, being in moments where time as an invention disappears. *How* is the question.

Concluding Questions

Larry: How can we live in the world of time and language day-to-day while recognizing that this idea of time is an invention? How do we address the issues of the world we live, like climate change, for example, as individuals and communities, such that biology is allowed to work? Do we act differently from how others are acting? How do we stay in the moment knowing that as soon as we step into the world of time and language we face a situation that appears already doomed to global disaster? How do we avoid or change the apocalyptic vision? Do we retreat to the moment? Do we stay optimistic about technological innovation? These are questions for the cybernetics community.

Jude: Sounds like a great set of questions for a cybernetic conference. As Pedretti said:

> These conferences each time regurgitate stuff and the next time it will be possible to say different things. The sentences in which these ideas were originally formulated were very contorted. And over time, they became easier to say … . So, the sense in which the language or the culture of our cybernetic language developed is the way in which our way of drafting things developed, is the way cybernetics developed. There is no distinction. I'll leave it there. (Pedretti in Lombardi, 2021, quotation starts at 48:55 min.)

References

Glasersfeld, E. von (2003). Adaptation, assimilation, accommodation. Video retrieved February 2022 from https://www.youtube.com/watch?v=LxTZTETSPOg.

Lombardi, J. (1996). Contextual essay for video entitled *From what to when is (not) violence*? UMI Microform 9625453, pp. 42–54. Ann Arbor, MI: UMI Dissertation Services.

Lombardi, J. (2021). *Annetta Pedretti: Turning objects into rhythms.* Video accessed February 2022 at https://www.youtube.com/watch?v=6Q78GHYle0s & www.youtube.com/watch?v=6Q78GHYle0s&t=2169s

Lombardi, J. & Maturana, H. (2021). The biology of love, language and languaging. Video accessed February 2022 at: https://www.youtube.com/watch?v=y4OqYmXA45g.

Lombardi, J., Werthamer, L., Cooper, J. (1993). A behavior continuum framework, prevention/intervention strategies and characteristics of an effective change agent. In *The Classroom Prevention Program Manual* (pp. 76–83). Baltimore, MD: Department of Mental Hygiene and Public Health, Johns Hopkins University.

Maturana, H. (1992a). *Structural determined entities*. Video accessed March 23, 2022 from https://www.youtube.com/watch?v=HuR-An8IOCw

Maturana, H. (1992b). Reality: The search for objectivity, or the quest for a compelling argument. In the conference workbook for "Language, Emotion, the Social, and the Ethical" (pp. 1–63). American Society for Cybernetics, Seabeck, Washington, July 9-13, 1992. Available in *The Irish Journal of Psychology*, 9(1), 25–82 (1988).

Maturana, H. (1995). The nature of time. Retrieved March 23, 2022 from https://sites.evergreen.edu/arunchandra/wp-content/uploads/sites/395/2018/05/natureOfTime.pdf

Maturana, H. (2008). Letter to the American Society for Cybernetics. Retrieved April 7, 2022 from https://asc-cybernetics.org/2008/HM-08WienerComments.pdf

Maturana, H. (2012). *Fundamental relativity*. Presentation at the 2012 Annual Meeting of the American Society for Cybernetics, July 9-13, 2012, Asilomar, California. Video accessed March 23, 2022 at https://www.youtube.com/watch?v=5WgpP0MfHXc.

Maturana, H. (n.d.). *Zero-time cybernetic*. Unpublished manuscript.

Richards, L. (2010). The anticommunication imperative. *Cybernetics & Human Knowing*, 17(1/2), 11–24.

Richards, L. (2016). A history of the history of cybernetics: An agenda for an everchanging present. *Cybernetics & Human Knowing*, 23(1), 42–49.

Richards, L. (2020). Acting cybernetically? *Cybernetics & Human Knowing*, 27(2), 17–26.

Seaton, N. (2021). *House of Annetta*. Video available at London School of Film. https://lfs.org.uk

Bunnell, P. (2019). *Distinguishing & Naming*. Photograph.

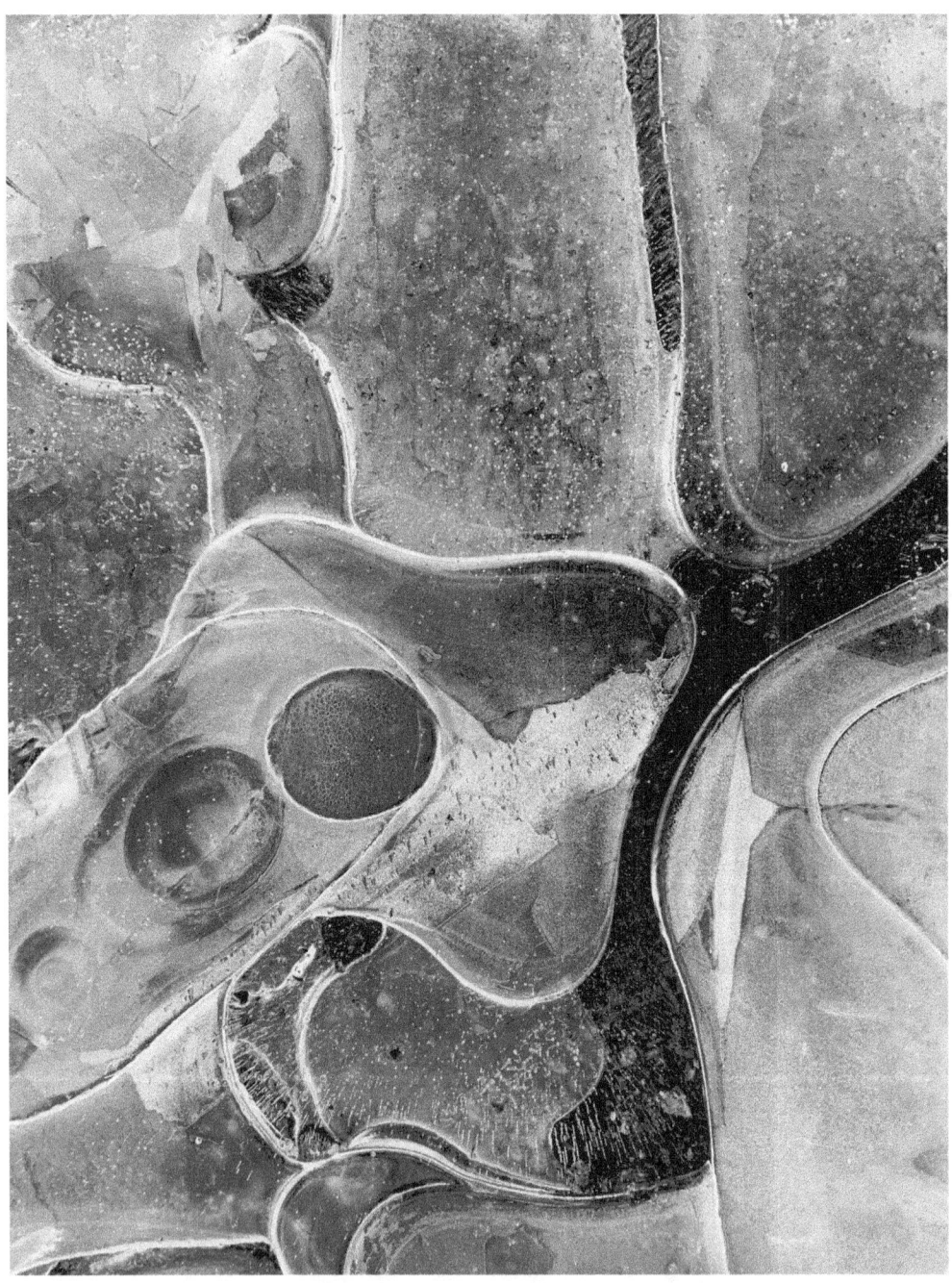

Bunnell, P. (2021). *Structural Coupling*. Photograph.

Animals and Humans Alike

Lloyd Fell[1]

> Maturana showed us how the biology of cognition concerns all living things. My research into the effects of stress on farm animals, prior to 1998, was transformed as I learned about autopoiesis and structural coupling. Since then, Maturana's teaching about love has captivated me, both in my teaching and writing.
>
> **Keywords:** Humberto Maturana, animal welfare, stress, love

When I first read about the work of Humberto Maturana, my research was aimed at improving the welfare, health and productivity of farm animals by reducing the impact of stress in their lives. During the 1970s and 1980s, I measured many physiological and behavioural responses to various stimuli that were potentially stressful.

In those days we emphasised a simplistic interpretation of the stimulus-response paradigm as a cause and effect. The dead weight of behaviourism still held sway, even though it was waning, and operant conditioning from repeated stimulus and response was the mainstream way of thinking about learning, certainly for animals. Deep down I was aware that this understanding of the living process was woefully inadequate. For one thing, each animal is different; what was stressful for one might not be for another.

Maturana inspired me to take a different approach; firstly, to stop regarding these animals as automatons that responded to each stimulus they encountered in a mechanical sort of way. I had grown up on a farm amongst animals and I respected them as living creatures like myself. But the term *mind* was taboo in animal science and there was no sign of a cognitive animal ethology until the late 1990s.

It was the idea of autopoiesis and structural coupling that changed the direction of my research. The individual responses to a stressful stimulus became a side issue. In the fundamental living process that Maturana described, they were merely *perturbations*, as he said often in his lectures. I learned ways to strengthen the resilience of animals by attending to their history, which includes training and habituation from an early age and having regard for their social grouping. This acknowledged their individual autonomy and also their connectedness.

My research was more successful and satisfying from then on, leading to gains in productivity and improvements in health in sheep and in cattle. These findings (which are reported in conference proceedings and industry reports) triggered changes in farm practice for rearing and weaning young cattle in Australia.

I was also concerned that the current methods of assessing the welfare of farm animals were inadequate because they did not recognise animal cognition (Fell, 1998).

---

1. Email: lloydfell77@gmail.com

I was reminding people that other animals are not some separate category of being. Maturana showed us how the biology of cognition concerns all living things.

On retirement, I could redirect my attention to what seemed to be even more important: the biology of stress in human society and the way it was linked to cognition and, indeed, to the biology of love. That became my intellectual passion and occupied all my time. I was lucky enough to have several University colleagues with whom to discuss Maturana's ideas and very fortunate that he came to Australia several times in the early 1990s for lecture series and discussions. My paper with David Russell in an early issue of this journal (Fell & Russell, 1994) set the scene for our many joint presentations, workshops and articles in which we tried to educate students and excite interest in these important ideas (e.g., Fell, Russell, & Stewart, 1994).

Since then I have self-published a series of books around the subject of mind and love.[2] These have less of the impeccability that I know Maturana valued, but they have been inspired by what I take to be his teaching about the biology of love. This is also celebrated in the book *Something Beyond Greatness*, in which his understanding of the systemic nature of our biological existence is linked to the experience of love.

> Greatness is not an achievement ... it appears as the observer speaks of it, when he or she sees some human being acting spontaneously with systemic vision and is surprised or moved ... What the observer has, in fact, seen is not more or less than the presence of love as the foundational emotion that moves, realises and conserves humanness. (Rogers & Naraine, 2009, p. 98)

The man whom I got to know during his several visits to Australia was very warm-hearted and gracious, and notwithstanding the forensic clarity of his explanations, a very humble person, especially considering what I regard as the profundity of his contribution to science. I will remember him with gratitude, love and respect.

References

Fell, L. (1998). Animal cognition in relation to farm animal welfare: The need for a different approach. *International Journal of Comparative Psychology, 11*(3), 1–21.

Fell, L. R., & Russell, D. B. (1994). Towards a biological explanation of human understanding. *Cybernetics and Human Knowing, 2*(4), 3–15.

Fell, L., Russell, D., & Stewart, A. (Eds.). (1994). *Seized by agreement, swamped by understanding: A collection of papers to celebrate the visit to Australia in 1994 by Humberto Maturana*. Sydney: Hawkesbury Printing, University of Western Sydney. Retrieved March 11, 2022 from www.biosong.org

Rogers, J., & Naraine, G. (2009). *Something beyond greatness*. Deerfield Beach, FL: Health Communications Inc.

---

2. These can be accessed at www.biosong.org

Power Arises Through Obedience
A Conversation About the Years of Dictatorship in Chile, the Helplessness of Power, and the Freedom of the Individual

Humberto Maturana and Bernhard Poerksen[1]

In this conversation with Bernhard Poerksen Humberto Maturana reflects about the years of dictatorship in Chile, his attempts to maintain the democratic spirit under extremely aggravated conditions, his efforts to support his students and an encounter with the dictator Pinochet in the presidential palace. The way and manner in which he describes his experiences and the drama of this encounter appear as an instructive demonstration of personal autonomy and systemic wisdom.

Keywords: autonomy, systems theory, power, oppression

Introduction by Bernhard Poerksen

Let me begin with a personal remark. Humberto Maturana and I first met in May 2000 in the rooms of the University of Chile in the centre of Santiago. There, in his laboratory, the plan was hatched that we produce a book together; one that would present his manner of thinking along the boundaries of science and philosophy in dialogical form. At this first meeting we managed to agree on the key topics which we discussed cautiously and hesitatingly, still groping for the proper form of expression on how we might approach the discovery of the human observer and the biology of knowledge. Unfortunately, torrential downpours that swamped half Santiago at that time made it impossible to meet often enough. In 2001 we were at last able to meet again in Santiago de Chile and I succeeded in recording daily conversations over several weeks, which I then edited and submitted to Humberto Maturana for adaptation and approval. The central concern in all these topically most diverse discussions and debates was the decisive change, the re-orientation, from being to doing, from the essence of an object to the process of its origins and development.

It became quite clear to me there that Humberto Maturana's approach was always a fundamental one, driven by enthusiasm and intellectual rigour, whether he dealt with the period of Chilean dictatorship, the education of children, or the theory of autopoiesis. What fascinated him was the investigation of the conditions forming a reality or even creating one. From such a perspective nothing can be seen as invariable or given once and for all. Everything can be referred back to, and explained from and through, its particular origins and development. In those conversations and the shared

---

1. Email: bernhard.poerksen@uni-tuebingen.de

subsequent public appearances in Hamburg, Berlin and Tübingen, Humberto Maturana imparted to me what science can be and do or in fact—ideally—should be: a mixture of rigorous and wild thinking at the interface of different disciplines. This definitely entails hard work on the one hand, but clearly also comprises an adventurous exploration through thinking and dialogue. Both of these aspects require enthusiasm and a huge portion of single-mindedness and personal independence. Humberto Maturana's exemplary conduct in this attitude inspired me profoundly and came to me at a most appropriate time. Namely I was in my early thirties when we first met and beginning to build my academic career. It encouraged me—and many others—to chart my own intellectual course.

For me, Humberto Maturana, who died on 6 May 2021, was a gentle and careful revolutionary who broke down the frontiers of normal science by his particular mixture of neurobiology and philosophy. I see his work as a path towards a *neurosophy* that is truly inter- and transdisciplinary. His concepts relating to the role of the observer in cognitive processes, his theory of autopoiesis and his thoughts about an anthropology of social life (his key concept was love) have thrilled, enriched and occasionally irritated the world of systemic thinking. However, I think that his greatest achievement is something else: Humberto Maturana has implanted the idea of autonomy in the centre of scientific and societal debate. He has spread the beautiful virus of self-sufficiency paired with resilience, celebrating individual autonomy, and so unveiled hitherto unrecognised degrees of freedom in thinking and practical living to his listeners all over the world. In this light I might add that autopoiesis, perhaps the best-known of Humberto Maturana's concepts, is the form, in which living systems realise their own autonomy.

For this issue of *Cybernetics & Human Knowing* I have chosen a chapter from our book *From Being to Doing*, in which Humberto Maturana talks about the years of dictatorship in Chile, his attempts to maintain the democratic spirit under extremely aggravated conditions, and his efforts to support his students. He also relates how he one day encountered the dictator Pinochet at a lunch party in the presidential palace. The way and manner in which he describes his experiences and the drama of this encounter appear to me to be instructive demonstrations of an ingenious way of rebelling against ideologies and repressive power. It shows, I think, the personal autonomy and the systemic wisdom of a great thinker.

The Emergence of Blind Spots

Poerksen: Let us look back to the years of dictatorship in Chile and the various efforts to survive in a totalitarian system while still preserving integrity and self-respect. Perhaps we could begin with the decisive historical moment, the day of the coup on 11 September 1973. At two o'clock in the afternoon troops of the putsching general Pinochet stormed the presidential palace—and at the end of the day Salvador Allende was dead and General Pinochet ruled the country as dictator for many years. Many members of the university fled to other countries and emigrated to the USA or to Europe. What did you do?

Maturana: On the day of the military coup I rang my friend Heinz von Foerster and asked him to help me and my family to leave the country. The situation was dangerous: Many people were suddenly among the persecuted, there were dead people lying in the streets, there was a curfew, and there were arrests. Soldiers appeared in the university. Heinz von Foerster tried to get me an invitation from an American university, which was not at all easy, of course. I was considered a dissident in science who spoke of the closure of the nervous system although everybody knew that it was demonstrably an open system. I was known but I did not belong to mainstream science. It was therefore not surprising that nobody wanted to have me at first, despite the efforts of Heinz von Foerster. The University of Illinois was not interested either. Ten days later, a neurophysiologist in New York had been found who was interested in my work. But by that time I had already decided to stay in Chile.

Poerksen: How did you reach that decision? There was an exodus of the intelligentsia in those years, an escape from repression and torture. Tens of thousands of Chileans emigrated, and the opposition was the object of unremitting persecution that some 3000 people did not survive.

Maturana: My motives to stay were of different kinds. My first thought was: If all democratically minded people left the country there would soon be no recollection of a democratic culture and of another, a better time. In this perspective, every older person was a living treasure. Then I was concerned about the fate of all the many students who were dispirited and suddenly found themselves drifting through the university on their own. Many professors had fled or gone into hiding, or had already been arrested. I met with some of them in the university one day, and we formed a sort of pact and decided to stay in Chile. I kept that pact and continued to work as a democratically minded member of my university because I felt responsible for the students and my country.

Poerksen: You wrote once that one of your motives was to comprehend the essence of dictatorship.

Maturana: That is true although it may sound a bit crazy. But I really wanted to know what it means to live under a dictator. I wanted to understand the Germans and, in particular, the history of my friend Heinz von Foerster who had survived the Nazi terror due to his capability of figuring out systems. He once said to me: "The more differentiated a system is the better you can cheat it." I also asked myself whether I might be able to observe in such a dictatorial system how people gradually go blind, and what the causes of such perceptual deprivation were. Can one, if one has been duly forewarned and is aware of the dangers of ideologically produced blindness, prevent it from developing and retain one's capabilities of vision and perception? One of the goals of a dictator is always to deprive people of any opportunity to remain or become observers of their own circumstances, and to deny them all chances of changing these circumstances and transforming them according to their own desires.

Poerksen: You wanted to come to grips with the epistemology of ideologies.

Maturana: You might put it that way, yes. When innumerable Germans insisted after the War that they had known nothing about the horrors of the Nazi period, I was convinced that not all of them were liars. Perhaps some of them were simply unable to face up to the terrible truth. I wanted to know what had been going on inside them and in their psyches. How does one live under a dictatorial regime that makes it so very difficult to keep away from it? In what measure does one unavoidably go blind even though one definitely does not want it to happen? Does one go blind because one knows that one could? How and in what ways is blindness produced at all?

Poerksen: What did you observe?

Maturana: Nobody is everywhere. If you decree curfews, you prevent people from seeing certain things. They will be unable to notice that people are murdered in their street during the night; they will not see the corpses. Everything happens behind a curtain. So people might not believe the rumours and tales they come across when they go out in the morning. There is nothing to be seen, not even a trace of blood, and what has happened is strictly denied and rejected by the authorities. Moreover, people will probably say to themselves that soldiers are human beings too, and that no human being can behave in such bestial ways. Such humanist presumptions may therefore very well make us blind: they protect us against the horror and they preserve our trust in other people. Of course, the new situation of a dictatorship creates new advantages for some people: Suddenly particular jobs are available because other people had to give them up and get away.

Ideology and the Military

Poerksen: Comparing the Chilean and the National Socialist dictatorships, we discover an essential difference: Adolf Hitler created an ideological dictatorship. He tried to win elections, although by applying massive means of intimidation, on the one hand, and he wanted to convince and enthuse the masses for his mad anti-Semitic ideas and the religion of racialism, on the other. The military dictatorship in Chile was primarily based on the force of arms and the power of the army; its ideological underpinnings were rather weak.

Maturana: That is a central point. The mental freedom of movement of people living under an ideological dictatorship is doubly limited: it is decreed, on the one hand, what has to be believed, and it is specified, on the other, what must never be said or thought if any risk of endangering life or status is to be avoided. A military dictatorship primarily lays down what must not be done. In the Chile of those years, any kind of criticism of the government and any sort of support for the ideals of socialism was prohibited. Apart from that, you could think and teach whatever you liked.

Poerksen: Pinochet kept reiterating that the Left was against the family, private property, freedom, and the fatherland. He used just a few meagre ideological phrases, nothing more.

Maturana: It was an *anti-ideology,* directed against communism. We were, after all, Pinochet kept pointing out all the time, in a state of war, and in a state of war you have to kill your enemies—that was his argument. He used this state of war declared by himself to justify the violations of human rights that were committed.

Poerksen: A central element of the Chilean dictatorial regime was the *miedo,* the terror, the spreading of fear. The singer and guitarist Victor Jarra was arrested, had his hands smashed, and was finally murdered. The poet Pablo Neruda was isolated; his houses were searched. People were tortured. Did people know about all this?

Maturana: Yes. For over a year every television newscast had to start with the bombardment of the government palace, then came reports about the arrest of revolutionaries and the discovery of secret arms caches. And so on. We should, however, not forget that Pinochet had the support of a significant majority of the population. Many people acquired enormous wealth under his regime through the privatisation of public property and, therefore, profited directly from the activities of his government.

Poerksen: I find it striking that you and various other authors, who are counted among the founders of constructivism today, all had to suffer under dictatorial regimes and were confronted with dogmatic worldviews. Heinz von Foerster had to hide from the NS-thugs; Ernst von Glasersfeld left Vienna when the Nazis seized power; Paul Watzlawick has repeatedly suggested how deeply shocked he was by the NS-regime; Francisco Varela escaped from Pinochet to Costa Rica, and you lived in Chile all those years. My question is now: Is there a connection between the theories of these authors and the experience of dictatorship? Alternatively, is this biographical correspondence purely accidental?

Maturana: It is not accidental but the result of the period. Infinitely many people were confronted with authoritarian systems more or less directly during the past century—the century of the Russian Revolution, of Fascism and National Socialism. I can, of course, only speak for myself, but my own understanding of power does not derive from the experiences I went through after the military coup in Chile. Rather the reverse. My life under the dictatorship was informed by my understanding of power, resulting from my permanent longing for democracy. Supporting democracy obviously entails the rejection of dictatorship that, therefore, becomes an enemy and a constant threat lurking in the background. All those actively engaged in the democratisation of a country quickly realise how difficult and laborious it is to keep a democratic culture alive. One has to come to terms with the ideal of perfection, which is widespread and deep-rooted in our culture, and with the attempt to generate seemingly perfect and allegedly democratic forms of living together even with the means of oppression. One is evidently opposed to dictatorship and, consequently, an active supporter of the individual, not of the goals of some collective. Still, one must not lose sight of the whole of society when working for the democratic participation of the individual. The persons you mentioned have, I think, been well aware of these difficulties and understood that there is no antagonism between individual and society. This is what they all have in common.

The Helplessness of Power

Poerksen: Your contributions to systems theory and the biology of cognition always deal with the autonomy of individuals and their particular ways of looking at the world and moving around in it. You claim that all human beings follow their own laws in cognition and action, that they are structure-determined systems. This conception sets narrow limits for the concept of direct and linear control. However, is not the wielding of power and force by dictators a compelling example of how extensively people can be controlled and influenced by external forces, after all?

Maturana: No, that is not the case. As I have lived under a dictatorial regime, I know what I am talking about. Strangely enough, power arises only when there is obedience. It is the consequence of an act of subjection depending on the decisions and the structure of the individuals subjecting themselves. It is granted to dictators by doing what they want. You grant power to others in order to keep or save something—life, freedom, possessions, jobs, a relationship, etc. I claim: Power arises through obedience. When dictators or other people point a gun at me and want to force me to do something, then I am the one who has to consider: Do I want to grant power to these people? Perhaps it is sensible to meet their demands for some time in order to be able to get the better of them in favourable circumstances.

Poerksen: Does what you are saying also apply to the dictatorship of the National Socialists? Was it the terror of the Gestapo that made Adolf Hitler powerful? Or did the people actually decide to grant power to a third-class painter from Austria?

Maturana: It was a conscious or a subconscious decision of the people, which gave power to Adolf Hitler. All those who did not protest had decided not to protest. They had decided to subject themselves. Suppose a dictator comes along and kills every person refusing to obey him. Suppose the people of the country all refuse to obey him. The consequence: He kills and kills. But for how long? Well, in the extreme case he will go on killing until everybody is dead. Where is the dictator's power then? It has vanished.

Poerksen: How do you want us to interpret this re-formulation of the relationship between power and helplessness? Is this an idealistic call not to subject ourselves? Or do you really mean what you are saying?

Maturana: I am totally serious when I say: We always do what we want to do, even though we may claim to be acting against our will or to have been compelled to do something. In such cases we desire the consequences of our actions although we may not like what we are doing at the moment.

Poerksen: Could you illustrate these ideas by an example?

Maturana: Nobody can force you to shoot at another person but you may, of course, decide to shoot in order to save your own life. Maintaining that you were forced to shoot is only an excuse that obscures the goal you were pursuing, namely, to save your life for the price of subjecting yourself. When you decide, in such a situation, not to shoot at another person, a shot may still be heard but it will be a shot fired at you—and you might die, preserving your dignity.

Poerksen: Would you say, therefore, that there are no real victims?

Maturana: Strictly speaking, yes. Victims despise themselves because they have granted power to others and denied themselves in their autonomy by an act of obedience. In the self-description as victims, the actual processes of the generation of power are made invisible.

Poerksen: The Chilean dictator Pinochet ordered, as we all know, the abduction, torture and murder of many of his opponents. What did you do when Salvador Allende was dead and the socialist experiment had met with a bloody end?

Maturana: I decided to pretend in order to stay alive and to protect my family and children. At the same time, I tried to move and behave in such a way as to avoid endangering my dignity and my self-respect. I kept away from certain situations, respected the curfew, did not discuss certain topics in the university. When the soldiers came and ordered me to raise my hands and to move up to the wall, I raised my hands and moved up to the wall. However, it was quite clear to me in those moments that the time would come when I would no longer be prepared to grant power to the dictator's regime.

Poerksen: Would you like to tell me about a particular situation?

Maturana: One day in the year 1977, I was arrested and put into prison. The reason was that I had given three lectures. The first lecture dealt with Genesis and the Fall. I said that Eve who had eaten from the apple and then given it to Adam could serve as an example. She was disobedient, and her rebellion against the divine commandment laid the foundation for human self-knowledge and responsible action, for the expulsion from paradise, a world without self-knowledge. In the second lecture, I spoke about St. Francis of Assisi. His way of perceiving human beings generates such deep respect towards them that it becomes impossible to define them as enemies. And I added that every army must first transform other human beings into strangers and then into enemies in order to be able to maltreat and kill them. The third lecture was devoted to Jesus and the New Testament. How do we live together, I asked my audience, if we base everything on the emotion of love?

Poerksen: What exactly happened after your last lecture?

Maturana: A few days later, I was taken to prison and treated like a prisoner. I was to be interrogated, I heard. One day somebody arrived and called out my name and said: "Are you Professor Humberto Maturana?" When I heard that I thought that I would remain a professor forever even if these people killed me. The status of professor was the protective shield they had granted me. They took me to a room where three persons were waiting. I sat down and asked the question: "In what way have I violated the statement of principles issued by the military government?" This means that it was me who began the interrogation and changed the rules of the game. I would not say that I manipulated those people but that the interrogation took place in a way that allowed me to keep my dignity and self-respect. I continued behaving like a professor and tried to counter the accusations they formulated. And I gave these people a lecture on evolutionary theory and explained to them why they would never be able to destroy communism by persecuting communists. It was necessary to change or eliminate the conditions that made communism possible, in the first place. The three men listened to me with growing astonishment. I told them they could invite me for a lecture any time. Then they took me back to the university.

The Maintenance of Self-Respect

Poerksen: Your very own experiences during the years of the dictatorship are most important to me because they make me understand you better, I believe. You do not plead for some fatal heroism, you do not condemn those who subject themselves, but you plead for a maximum of awareness in the handling of power.
Maturana: Naturally, yes. It can be very stupid not to subject oneself for a time and to wait for a suitable opportunity to strike back. My fundamental point is to declare one's responsibility and to invite others to act in full awareness. Does one want the world that emerges when one grants power to others? Does one primarily want to survive? Does one reject the world emerging through the wielding of power in an unconditional and uncompromising way?
Poerksen: Do you believe that that different state of awareness is decisive? It might be argued that conscious or subconscious subjection leads to the same consequences: the dictator stays in power.
Maturana: This different state of awareness is decisive because it allows you to pretend. Pretending means simulating a non-existent emotion. You remain an observer, keeping an inner distance, and one day you may act in a different way again. This means that the perceptual abilities of the pretenders are not destroyed, and their self-respect and dignity are preserved. Due to these decisive and very significant experiences, they may be able to lead a different life. If one gives up this attitude of the conscious handling of power, one is lost because one has decided for blindness.
Poerksen: How can we be sure that the belief that we are merely pretending and observing is not just a subtle and refined form of self-delusion?
Maturana: Well, that is a difficult problem, indeed. The situation is particularly precarious when people are convinced that they are immune to the temptations of power. These people have become blind to their own temptability, to the delights of wielding power, the pleasures of the uncontrolled execution of control. My view is that we should never believe that we are in any way special as far as morality or anything else is concerned: we are then mentally unprepared for situations that may make torturers of us. Those who think they are immune will be the first, I believe, to become torturers in certain situations. They are not aware of their own seducibility. Whatever horrible or wonderful things one human being can do—there will always be another, and it could be you or me, who is capable of doing the same. Such an insight allows us to lead our lives in full awareness and to decide whether to support democracy or a dictatorship.
Poerksen: Throughout the 17 years of the dictatorial regime in Chile, you worked as an academic teacher together with your students. How openly could you operate within the university? How did you conduct your courses?

Maturana: It was still in 1973 that I invented a series of lectures entitled *Biology of Cognition*, which later became the book *The Tree of Knowledge*. I gave these lectures year after year, describing the way from the single cell to the social. I was careful not to attack the government in any direct way or to campaign openly for some political end—that was not my thing. I never urged my students to go in a certain direction but I wanted to develop their capacity for reflection step by step.

Poerksen: If I understand correctly, you wanted to teach them how to think independently. Could you present an example of your teaching to illustrate your approach?

Maturana: I once talked about my view that power is granted through obedience. Nobody possesses power, I said, but they are given power by others who subject themselves and do what is demanded of them. I had come to the lecture with a very realistic toy gun. "With this gun," I said to my students, "I can kill you." I pointed at a female student and said: "Stand up, or I will shoot you!" She stood up although she knew, of course, that I would never shoot her. "Come to the centre of the room!" She went to the centre of the room. "Lie down on the floor!" She lay on the floor. "Take off your clothes!" In this moment she jumped up and shouted: "No! That I will not do!" I waited for a moment and then said: "You see, this refusal to obey has robbed me of my power. My power rests on your willingness to obey and not on the fact that I am waving a gun about." You see, I did not tell my students what they had to do, but I tried to lead them to other possibilities of reflection and perception. My view is: Those who favour a certain way of living and desire that that way of living arise and reveal itself in the relation to themselves, should live it without hesitation. Waiting will not be of any use.

Poerksen: Structure-determined systems—human beings—can only be controlled in a limited way; one can perturb them but not control them. Compulsion appears to stand no chance, in principle. My thesis is: You have developed an epistemology that removes the conceptual foundation of dictatorial power.

Maturana: I strongly support this thesis and want to add that I can destroy the conceptual foundations of dictatorship because my work allows me to achieve a more profound understanding of democracy. Democracy must be created anew every day, I believe, as a space of living together in which participation and cooperation are possible, based on self-respect and the respect of others. The first thing a dictatorship destroys is the self-respect and the autonomy of every single individual, because it demands subjection and obedience as the price for staying alive.

Poerksen: Could it be that the immense popularity of your ideas today is due to the often-invoked end of all ideologies and the collapse of the sort of socialism that really existed?

Maturana: I see a connection. What I have written provides a new foundation for the possibility of self-respect, which is fundamentally negated by dictatorships. What the readers of my work may realise is that we are all unavoidably participating in the creation of the world we live in. This is the view that I invite people to try without compulsion or cost, a view that values the individual. And whoever feels appreciated and respected, will be enabled to appreciate and respect themselves. They can accept the responsibility for what they do.

Encounter with Pinochet

Poerksen: I have been told that you once actually met the dictator Pinochet. Would you like to tell me about the circumstances of this encounter?
Maturana: One day, it was the year 1984, I received a letter with the seal of the president. It was an invitation to have lunch with Pinochet, which had also been sent to other members of the faculty, as I found out later. Some people thought we could not decline, others warned us not to attend the dinner, but I decided to accept the invitation. My mother implored me to remember all the time that I had a family, and I promised her not to forget that. When I finally arrived at the presidential palace, I found that about 85 professors had assembled there. We stood around for a while, talked to each other, and asked ourselves why we had been invited at all. Then Pinochet appeared. His attendant told him the names while he welcomed us. When my turn had come to greet him, I thought of my eldest son who had said to me that he would never shake Pinochet's hand. And there I was, and shook this man's hand. After that, we went to eat in a vast and magnificently decorated hall. As soon as we had sat down, Pinochet rose again, took his wine glass, and said: "Let us drink to our fatherland!" And we rose, drank to each other, sat down again, and ate the delicious meal that was served on elegant porcelain specially manufactured for the President of the Republic.
Poerksen: You sat there with a man who ran a secret police that spread fear and terror, who was responsible for the disappearance without trace of numerous critics of the government, and who ordered people to be tortured.
Maturana: That is what it was like, precisely. Before dessert was served, Pinochet, who was sitting only a few metres away from me, addressed us again. "Ladies and gentlemen," I heard him say, "the sole purpose of this meeting is to get to know each other. That is all. You may feel quite safe; there will be no demands on you of any kind." He sat down again, and in that moment I picked up my glass, stood up and said: "Ladies and gentlemen, I would also like to toast our fatherland with you!" There was dead silence instantly. One could sense the deep alarm of the assembled persons, who seemed petrified with sudden fear. Pinochet looked at me and leaned forward a little. "We are gathered here today in the company of the president," I went on. "And that is a rare occasion under any government. I will, therefore, seize the opportunity and bring out a toast with you and the president to the effect that we all who are here today contribute to the intellectual freedom and the cultural autonomy of our country, Chile." I drank my wine; Pinochet leant back and clapped his hands four times. All the people in the room clapped four times. One of my friends turned to me and whispered: "Many thanks, that was wonderful." And general talking began again.
Poerksen: The dictator did not comprehend what you said.

Maturana: Just a moment, please, the story is not finished yet. Shortly after the dessert was eaten, we all went to another room. A friend of mine, a physicist of our university, pointed out to me that Pinochet was alone and that we should join him. I did not want to at first but he urged me on and so I finally went with him to join Pinochet who was standing there with one of his generals. "Mister President," my friend said, "I have the pleasure to introduce to you Professor Maturana, a very renowned biologist." I shook his hand again and he said: "I share your good wishes for this country." "*A dios rogando,*" I answered, "*y con el mazo dando.*" This is a Spanish proverb and means roughly: If you pray to God for something you must also act accordingly; prayers and pious wishes are not enough. It really was a bizarre situation: Pinochet was standing there and telling me that he shared my desire for intellectual freedom and cultural autonomy. All the goals of his politics were the direct opposite. He wanted to make this country dependent on others in order to be able to crush the first sproutings of communism with the help of his allies.

Poerksen: You spoke with a man who was thought to be rather limited by many people. Salvador Allende, who had put Pinochet in the position of power, in the first place, from which he could venture his putsch, once said that he considered him "too dumb to deceive his own wife."

Maturana: That was a crass misjudgement. Nobody is made an army general anywhere in the world if he lacks the necessary intelligence. He may be fanatical, narrow-minded and ideological—but he is not stupid.

Poerksen: What do you think? How did Pinochet understand what you said?

Maturana: He understood me perfectly well. The essential thing was that I did not treat him as a superior but as an equal Chilean. He was the president for me, he went along with us, and he had to contribute to this grand task of guarding intellectual freedom and cultural autonomy in the country. He was one of us, and that was not meant to be an insult, not at all.

Poerksen: You re-interpreted the relation between the ruler and his subjects.

Maturana: One could put it this way—and, furthermore, I used the words he had used in his toast: I also drank to our common fatherland.

Poerksen: I find this very revealing. You used the eigenlogic of a closed system in order to invade and transform it. You knew, of course, that *fatherland* was an excellent word for that.

Maturana: Quite so. You cannot, of course, impress an Adolf Hitler with an after dinner address in which you talk about the Jews and call for their veneration. One must also see clearly that insults cannot be successful in such situations. Whoever does not see and understand that is completely blind.

Poerksen: This implies, however, that one can exploit the eigenlogic of a system in a subversive way, to put it more generally.

Maturana: The orientation towards the eigenlogic of the system will work only as long as the meaning or the re-interpretation of what is said cannot be understood as a devaluation of the system. An insult (such as: "You are just a lousy dictator!") would, of course, be quite idiotic because Pinochet would have had to react to it. I was, therefore, extremely careful not to provoke him in any way but to appeal to a common vision: He could not possibly object to a plea for an effort in the service for our beloved country.

Poerksen: How did that encounter end?

Maturana: While we were still talking, another scientist approached and addressed Pinochet in an extremely servile manner. Pinochet stood to attention at once, became the dictator again, and answered brusquely: "What do you want?" I did not want to be associated with this form of servility and withdrew. When Pinochet turned to leave he came my way again, touched my arm and said: "Chao." And I said: "Chao!" I would say that he treated me as a Chilean of equal status because—without being arrogant—I had not subjected myself to him and had not given him power.

Poerksen: Did you ever meet again?

Maturana: No, never. In the evening of the same day, I received two kinds of telephone call: Some people were beside themselves with fury because they thought I had put everyone at risk; others thanked me. One of the professorial colleagues said the wording of my toast had given them back their dignity.

Poerksen: I am quite touched by this experience because it shows that there are always degrees of freedom, behavioural slots, which may be exploited by individuals in different ways. I am sure, however, that such behaviour as yours necessarily depends on talent and intelligence.

Maturana: Such behaviour has nothing to do with intelligence, certainly not. What you need perhaps is a good measure of wisdom based on a capacity of perceiving without prejudices and presumptions. If you approach such a dictator with the image of a terrible idiot and a criminal filling your mind you will inevitably behave in a particular way. Of course, that man is a criminal, no doubt about it. And, of course, he appears completely blind to his responsibility for what happens in Chile and for the horrors of his dictatorial regime—as we can tell by his speeches. But if we cling to this assessment we will not be able to see the human being in his prison, with his mental conflicts, and with his patriotism that is, after all, responsibly intended, and to address this human being when talking to him.

Poerksen: The years of the dictatorship are now definitely gone. In the year 1989 free elections were held again in Chile; the country now struggles with the problem of an adequate evaluation of its past. If another opportunity should arise to meet Pinochet, who is now an internationally stigmatised old man—although still revered by many Chileans—what would you tell him?

Maturana: I would advise him to act like Bernardo O'Higgins, the great Chilean freedom fighter. When he was accused publicly one day to have changed into a tyrant, he answered the enraged populace: "Whatever I have done—I have done it with the conviction that it would be beneficial to our country. If the pain and suffering that I may have caused can be relieved by giving my blood then I am prepared to die." Ultimately, O'Higgins was not killed but went into exile in 1823. He was prepared to assume responsibility for his actions and to succumb to the judgment of others.

Pinochet has never done that. He still insists that he is innocent. That is his greatest crime.

Acknowledgment

This conversation is an excerpt from the book: *From Being to Doing. The Origins of the Biology of Cognition* by Humberto Maturana and Bernhard Poerksen, published by Carl-Auer and translated by Wolfram Karl Koeck and Alison Rosemary Koeck. For more information see: http://www.carl-auer.com

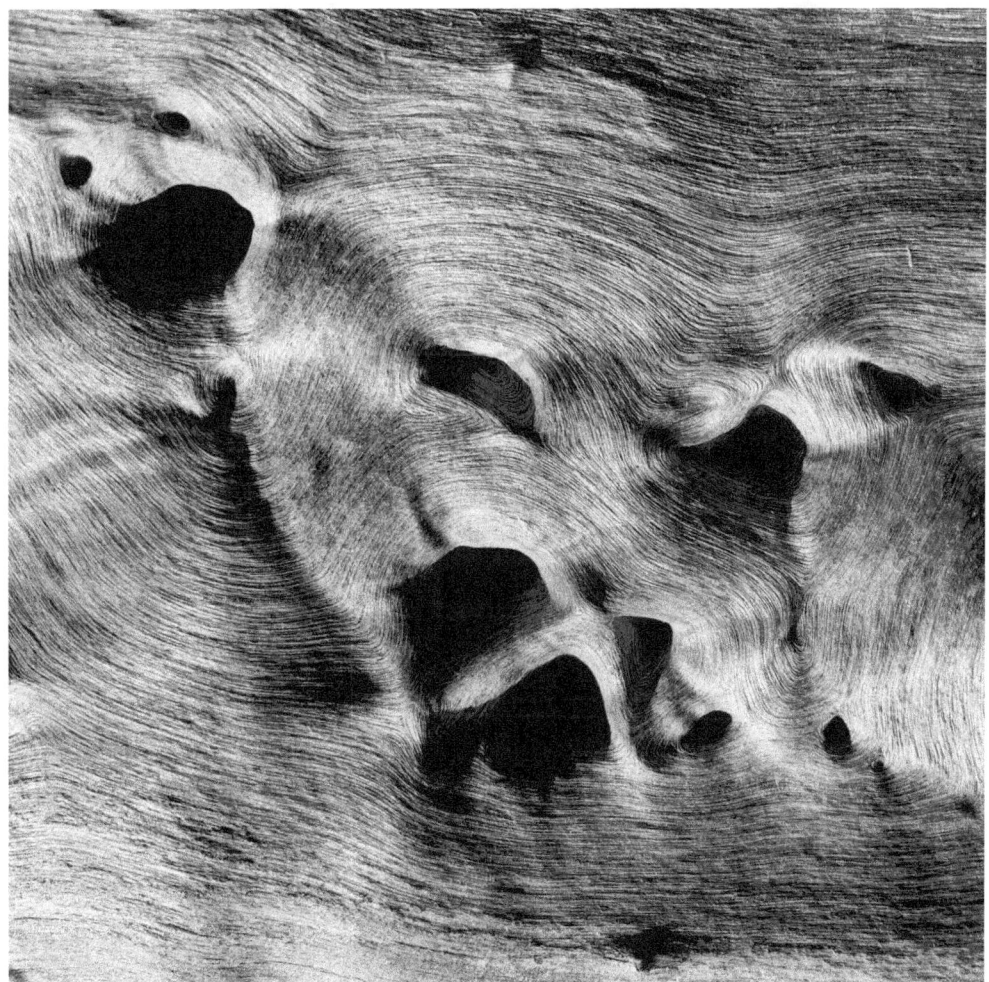

Bunnell, P. (2018). *Conservation & Change*. Photograph.

Bunnell, P. (2010). *Observer Observing*. Photograph.

Second Order Cybernetics and the End and Beginning of Philosophy

Bernard Scott[1]

> "But the wisdom that comes from heaven is first of all pure; then peace-loving, considerate, submissive, full of mercy and good fruit, impartial and sincere."
> — James, 3:17

Introduction

In an earlier guest column, entitled "In Defence of Pure Cybernetics" (Scott, 2019), I argued that cybernetics, as a trans-discipline and meta-discipline concerned with control and communication, did not need to be augmented by other disciplines or alien paradigms. As examples of these, I cited phenomenology, semiotics and psychoanalysis. I also pointed out that within the social sciences there are examples of process-oriented approaches that are compatible with cybernetics and, hence, can be usefully absorbed into cybernetic thinking. As examples, I mentioned the pragmaticism of C. S. Peirce, the social behaviourism of G. H. Mead and the psychological and educational theories of John Dewey. In the present article, I take my thoughts about the role of cybernetics a step further and argue that, with the advent of second order cybernetics, the traditional discourses of philosophy became redundant and are now of historical interest only. The exceptions to this are metaphilosophies, which address the epistemological status of philosophical problems and the nature of philosophy itself. Such metaphilosophies are cognate with second order cybernetics. Particularly relevant, is the recent work of Peter Hacker, who uses a Wittgenstein approach to *philosophical anthropology*, the study of the human condition, which takes into account the findings of empirical studies in the biological and social sciences, together with arguments about how those findings should be interpreted. (I say more about Hacker's work later). Given the conceptual ground clearing and clarifications that metaphilosophies and second order cybernetics provide, I propose that philosophy now has the opportunity for a new beginning by returning to its ancient roots as the *love of wisdom* (Greek, *philo*, loving, and *sophia*, knowledge, wisdom).

Second Order Cybernetics

I take cybernetics to be the science of control and communication (Wiener, 1948). It can also be practised as an art, the art of assuring the efficacy of action (Couffignal,

---

1. Email: bernces1@gmail.com

1956). The scope of cybernetics from its birth included man-made systems, biological systems and social systems. Towards the end of the 1960s and beginning of the 1970s, several thinkers helped develop an architecture for cybernetics, which hinged on the distinction between first order and second order cybernetics, where first order cybernetics is the cybernetics of observed systems and second order cybernetics as the cybernetics of observing systems. Heinz Von Foerster articulated this distinction (Von Foerster et al., 1974; see p. 1). It was similar to the distinction made by Gordon Pask between taciturn and language oriented systems (Pask, 1969). In later writings, Pask referred to the *new cybernetics* or *neo-classical cybernetics*, within which he placed his own cybernetic theory of conversations (Pask, 1975). It was also similar to Humberto Maturana's theses about the biology of cognition and his assertion that, "Anything said is said by an observer" (Maturana, 1970, p. 4), and the ecology-of-mind concept of Gregory Bateson (1972).[2] Here, I am using the label *second order cybernetics* to adumbrate all these approaches.

A useful way of characterising these developments is to see them as having a concern with epistemology and, in particular, the philosophy of science. What are focused on are the control and communication processes that take place within human communities, that is, communities of *observers*, systems that not only observe but also reflect on their observations and their conversations with themselves and with others. This communication about communication, control of control, and reflection about reflection, is what makes cybernetics a second order discipline. It immediately offers itself as a rational framework for addressing the physical sciences, the social sciences and the humanities. Cybernetics is a useful tool for improving the clarity of communication, a tool that can be used to make sense of the noise and redundancy to be found in the marketplace of ideas.

To appreciate the significance of cybernetics and, in particular, second order cybernetics, one needs an understanding of its key concepts. I have set these out in a recent book (Scott, 2021). Here, I wish to highlight one of them, the concept of organisational closure or, equivalently, autopoiesis (self-creation). As terms, both refer to dynamically self-organising systems whose organisation is such that the organisation reproduces itself from moment to moment.[3] In the terminology of Humberto Maturana, the structure of such a system may change but, for the system to survive, the circularity of its organisation has to be conserved, otherwise, the system will cease to be viable (Maturana & Varela, 1980). For observers and collectives of observers (persons, societies, cultures), one can distinguish between the biological system and its material augmentations (clothing, tools, vehicles ...), and the psychosocial system which is embodied by the biological system. The integrity of the biological and psychosocial are both required for viability.

The critical point to appreciate is that organisationally closed systems, as best they can, adapt to the perturbations they encounter within their environments. In doing so,

---

2. I provide an historical introduction to second order cybernetics in Scott (2004).
3. "Autopoiesis is that organization which computes its own organization"; "Autopoietic systems are thermodynamically open but organizationally closed" (Von Foerster, 2003, p. 281).

they construct expectations about future encounters and how to respond to them. Equivalent terminology includes, *becoming in-formed* (constructing a form within), *constructing a model, acquiring habits, forming concepts, acquiring eigenbehaviours, constructing universals, distinguishing objects*, and *anticipating*. For observers, this implies that the meaning of the message lies with the receiver. Each observer constructs her own reality. It then behooves her to take responsibility for what she has constructed. Thus, second order cybernetics provides a conceptual framework for addressing questions about affect (emotions), cognition (mind), and ethics (the "good" life) in a unified way. It avoids the conceptual confusions found in modern day versions of materialism and reductionism. Peter Hacker's work, mentioned earlier also targets these conceptual confusions. His ideas are set out in a series of books. The *categorial framework* (his term) that he develops is similar to that developed in second order cybernetics (see Hacker, 2010). However, as is the case with many thinkers who discuss the human condition and such topics as consciousness and the self, he shows no understanding or awareness of second order cybernetics.

Philosophy

The label *philosophy*, traditionally translated from the Greek as the love of wisdom, has been used in a variety of ways to characterise human intellectual endeavours. In the Western world, numerous parts and subcategories have been distinguished, for example, epistemology, ontology, ethics, aesthetics, logic and metaphysics. Over many centuries, other domains have been distinguished: natural philosophy begat science, disciplines that deal with number, shape and dynamics evolved into the complex labyrinths of modern mathematics (Greek, *mathemos*, that which is to be taught), philosophy of mind begat cognitive psychology, and so on. Meanwhile, the newly distinguished disciplines and domains of practice acquired their own adjunct, higher-order disciplines: philosophy of science, philosophy of psychology, philosophy of mathematics, philosophy of logic, and many more, so many that the label *philosophy* has all but lost any clear meaning in common discourse.

Running as a thread from the very beginnings of Western philosophy, there has also been discourse that implicitly and, at times explicitly, addresses the *philosophy of philosophy*. In its simplest form, this can be found in the introductions to standard textbooks of philosophy, which attempt to say what philosophy is about.[4] In a more carefully articulated form, it can be found in the philosophical investigations of Ludwig Wittgenstein and his followers, who look carefully at how language (communication with oneself and others) is used, such that humans become puzzled about questions that have been thrown up by the use of language itself.

> Philosophical problems ... are, of course, not empirical problems; but they are solved through an insight into the workings of our language, and that in such a way that these workings are recognized —despite an urge to misunderstand them. The problems are solved, not through the contribution of

---

4. As an example of this genre, I recommend Roger Scruton's (1994) *Modern Philosophy.*

new knowledge, rather through the arrangement of things long familiar. Philosophy is a struggle against the bewitchment (Verhexung) of our understanding by the resources of our language. (Wittgenstein, 1967, p. 47)

Wittgenstein developed a number of concepts about how language is used in what he refers to as language games, and showed that a large class of philosophical problems are pseudo-problems. A classic example is the unthinking belief that because there is a word for something that there is an essence or essential meaning that can be analysed and defined. Examples include the questions: What is mind, what is being, what is free will, what is truth, what is justice, what is the good, is there a deity, and, of course, does life have a meaning and, if so, what is it.[5] One should not ask what a word means. Rather, one should look to see how it is used. As pointed out by Humberto Maturana, a philosopher, after a preliminary outline of the problem, adopts a point of view and argues for it, whilst at the same time criticising the views of those who disagree with her. In contrast, Maturana characterises his own approach as that of a scientist who argues from evidence and uses logic to propose falsifiable hypotheses.

> My concern is with the scientific explanation of the biological phenomenon of cognition not with reality ... This is not a trivial difference because it reveals the differences between scientific and philosophical approaches in the processes of explaining or generating understanding. In the philosophical approach, the philosopher generates an explanatory theory with the desire to conserve certain principles or notions that he or she deems to be valid in themselves. (Maturana, 1991, p. 387)

As a shorthand, I like to refer to the approach of Wittgenstein as metaphilosophy (philosophy of philosophy). Wittgenstein characterises his analyses of language use to dispel philosophical pseudo-problems as showing a fly the way out of a bottle. Wittgenstein's metaphilosophy entails communication about communication. As such, it can be readily incorporated into cybernetics as a cybernetic practice. I advocate this move as one that helps tidy up and clarify the whole business of humans communicating with and about humans. Cybernetics is by its very nature critical cybernetics. It uses the tools of good science (evidence and logic) to help us agree about what is the human condition in all observable contexts, from the global level down to the level of individual well-being or lack of it.

As a case study of how second order cybernetics puts an end to centuries of philosophical wrangling, in this case, about the self and self-consciousness, I cite Von Foerster's analysis, using the mathematical theory of recursive functions, of the sensorimotor system as a circular process: Sensation is a function of motor activity and motor activity is a function of changes in sensation. Regularities (eigenbehaviours) are distinguished and labelled as objects and events. At the base of the recursion, "An organism is its own ultimate object" (Von Foerster, 2003, p. 256).

Von Foerster also uses recursive function theory in his analysis of the role of the reflexive pronoun, *I*, as a label to denote self-observation: "I is the relation between

---

5. For more philosophical problems, see https://en.wikipedia.org/wiki/List_of_unsolved_problems_in_philosophy (accessed February 7th, 2022).

self and observation of self" (Von Foerster, 2003, p. 257), and in his account of orders of consciousness and self-consciousness ("I observe me observing me observing me observing me ..."), which he summarises in the aphorism, "I is a relator of infinite order" Von Foerster, 2003, p. 257).

Interestingly, there are parallels in Wittgenstein's discussions of the grammars of self-reference and consciousness, usefully summarised in Scruton (1994, pp. 484–493). Scruton goes on to say, in a section about the social construction of the self, "Communities are not formed through the fusion of the agreement of rational individuals: It is rational individuals who are formed by communities. Membership comes first, and is the precondition of the outlook that would reject it" (p. 494). Von Foerster, as early as 1973, used a number of empirical and formal arguments to conclude that "Reality = Community" (Von Foerster, 2003, p. 227).

With this conclusion, Von Foerster proposed two imperatives: *"The ethical imperative:* Act always so as to increase the number of choices. *The aesthetical imperative*: If you desire to see, learn how to act." (Von Foerster, 2003, p. 227)

The first is a corollary of Ross Ashby's law of requisite variety and can be seen as a call for members of community to not impose unnecessary constraints on themselves. The second is based on the understanding that, within the circularity of the sensorimotor system, seeing is always a consequence of action. Elsewhere, Von Foerster (2003, p. 206) proposes a third imperative: "Act towards the future you desire."

The End of Philosophy

In what I regard as a very useful and succinct formulation, Heinz Von Foerster, drawing on the metaphilosophy of Wittgenstein, has made a distinction between decidable and undecidable questions (Von Foerster, 1997, 2003). The former are questions that can be resolved by empirical means or formal means (calculation). The latter cannot. The answers, or lack of them, are in the gift of the observer. The observer is free to choose the answer to an undecidable question. A ready to hand example of an undecidable question, regularly asked in the Roman/Graeco/Judaic world, is whether or not there is a God and/or some other "supernatural" beings. Possible answers are many and may be articulated in a bewilderingly large number of ways, as forms of theism, atheism and agnosticism. Another ready to hand example of an undecidable question is the question of what happened before the big bang (assuming one subscribes to the theory of the universe beginning with a big bang). As Von Foerster neatly points out, the question is undecidable because no one can ever be in a position to observe anything prior to the big bang.[6] However, with the freedom to

---

6. It is a sad comment on modern science that much of it consists of speculations about undecidable questions. Indeed, within the community of physicists, there is a group (the Quantum Bicycle Society) who argue for a return to the requirement that hypotheses should be falsifiable, that is, open to empirical test. See https://quicycle.com/ (accessed January 21st, 2022).

choose the answers to undecidable questions goes the responsibility for the consequences of making those choices.

Von Foerster, the cybernetician, has applied Occam's razor, the dictum that categories should not be multiplied unnecessarily, with remarkable effect. With the distinction between decidable and undecidable questions, traditional philosophical questioning comes to an end. The large canon of work that has been generated and, indeed, continues to be generated, is now of historical interest only.

Interestingly, in his essay, "The End of Philosophy and the Task of Thinking," the phenomenologist philosopher, Martin Heidegger (1978, p. 434), declares, "No prophecy is necessary to recognise that the sciences now establishing themselves will soon be steered by the new fundamental science of cybernetics." He goes on to say:

> This science corresponds to the determination of man as an acting social being. For it is the theory of the steering of the possible planning and arrangement of human labour. Cybernetics transforms language into the exchange of news. The arts become regulated-regulating instruments of information.
> The development of philosophy into the independent sciences that, however, interdependently communicate among themselves ever more markedly, is the legitimate completion of philosophy. Philosophy is ending in the present age. It has found itself in the scientific attitude of socially active humanity. But the fundamental characteristic of this scientific attitude is its cybernetic, that is, technological character." (Heidegger, 1978, p. 434).[7]

Later in the essay, he says,

> Perhaps there is a thinking which is more sober minded than the incessant frenzy of rationalisation and the intoxicating quality of cybernetics ... a thinking outside of the rational and the irrational... The task of thinking would then be the surrender of previous thinking to the determination of the matter for thinking" (Heidegger, 1978, p. 449).

It is clear from these statements, first made in 1966, that Heidegger had in mind first order cybernetics, the cybernetics of observed systems. My reading is that, in questioning cybernetics, Heidegger was doing cybernetics of cybernetics, that is, second order cybernetics. Thus, as Von Foerster, Pask, Maturana, and others have pointed out, the inescapable matter for thinking is the question of how to behave responsibly and ethically, that is, wisely, in a troubled world.

A New Beginning for Philosophy

I propose that philosophy now has the opportunity for a new beginning and a fresh identity by returning to its ancient roots as the love of wisdom (Old English, *wis*, learned, experienced, crafty, and *dom*, judgment). Outside the narrow confines of Western scientific thinking and the canon of traditional Western philosophy, there is a wealth of literature and art (stories, parables, aphorisms, images ...) that have been

---

7. Today, the technological face of cybernetics can be found in A.I., robotics and management cybernetics. The latter, of course, necessarily includes, explicitly or implicitly, second order cybernetics in its concerns with the behaviour of human participants (see Scott, 1997).

constructed in order to preserve and communicate wisdom, that is, to have an educative function by providing insight, understanding and enlightenment about the human condition (I have in mind the Latin root, *e-duco*, to lead out, by implication, to lead out from darkness into the light and from ignorance into knowing). As I write, I am very aware that the words I am using are not as well defined as are scientific concepts that use models and operational definitions. *Insight* speaks to the processes of seeing within; *understanding* speaks to the processes of explaining concepts in terms of other concepts, where *to explain* means to lay things out clearly, as on a plane surface; understanding can also refer to having compassion for (sharing feelings with and caring about) fellow humans; *enlightenment* is as used in English, where the word *lighten* has two distinct uses: to mean illuminating or bringing into the light, and/ or to mean lessening a load. I see wisdom itself as the judicious combination of love (of life and of one's neighbour) and understanding. I recommend that all observers take it upon themselves to study wise teachings and to help disseminate them. In this pursuit, reason is a tool only. Scholarly disputations are of minor interest. Obsessing about them is a sign of poor practice. My proposed new beginning for Western philosophy universalises the idea that all of us are cybernetician-philosophers, developing and applying the art and science of fostering goodwill.[8]

"Be wise as serpents and innocent as doves," (Matt, 10:16).

References

Bateson, G. (1972). *Steps to an ecology of mind*. New York: Chandler Publishing.
Couffignal, L. (1958). Essai d'une définition générale de la cybernétique. *Proceedings of the First International Congress on Cybernetics* (pp. 46–54), Namur, Belgium, June 26–29, 1956. Pairis: Gauthier-Villars.
Hacker, P. M. S. (2010). *Human nature: The categorial framework*. London: Wiley-Blackwell.
Heidegger, M. (1978). The end of philosophy and the task of thinking. In D. F. Krell (Ed.), *Martin Heidegger: Basic writings* (pp. 427-449). London: Routledge.
The Holy Bible, English Standard Version Anglicised (2001). Wheaton, IL: Crossway.
Maturana, H. (1970). Neurophysiology of cognition. In P. L. Garvin (Ed.), *Cognition: A multiple view* (pp. 3–24). New York: Spartan Books.
Maturana, H. R. (1991). Response to Jim Birch. *Journal of Family Therapy, 13*(4), 375–393.
Maturana, H. R., & Varela, F. J. (1980). *Autopoiesis and cognition*. Dordrecht, The Netherlands: D. Reidel.
Pask, G. (1969a). The meaning of cybernetics in the behavioural sciences. In J. Rose (Ed.), *Progress of cybernetics. Vol. 1* (pp. 15–45). London: Gordon and Breach.
Pask, G. (1975). *Conversation, cognition and learning*. Amsterdam: Elsevier.
Scott, B. (1997). Inadvertent pathologies of communication in human systems. *Kybernetes, 26* (6/7), 824–836.
Scott, B. (2004). Second order cybernetics: An historical introduction. *Kybernetes, 33* (9/10), 1365–1378.
Scott, B. (2019). In defence of pure cybernetics. *Cybernetics and Human Knowing, 26*(4), pp. 99–109.
Scott, B. (2021). *Cybernetics for the social sciences*. Boston: Brill.
Scruton, R. (1994). *Modern philosophy*. London: Sinclair-Stevenson.
Von Foerster, H. et al. (Eds.) (1974). *Cybernetics of cybernetics*. BCL Report 73.38, Biological Computer Laboratory, Dept. of Electrical Engineering, Urbana, IL: University of Illinois.
Von Foerster, H. (1997). Lethology: A theory of learning and knowing vis à vis undeterminables, undecidables and unknowables. *Revistas Universidad EAFIT, 33*(107), 13–30. Retrieved January 31, 2022 from https://publicaciones.eafit.edu.co/index.php/revista-universidad-eafit/article/view/1121/1012
Von Foerster, H. (2003). *Understanding understanding: Essays on cybernetics and cognition*. New York: Springer. Retrieved January 30, 2022 from https://www.alice.id.tue.nl/references/foerster-2003.pdf
Wiener, N. (1948). *Cybernetics*. Cambridge, MA: The MIT Press.
Wittgenstein, L. (1967). *Philosophical investigations* (3rd ed.). Oxford: Basil Blackwell.

---

8. Some useful starting points for the study of wisdom are *Proverbs* and *Ecclesiastes* in the *Old Testament*, the teachings of Jesus and the letters of the apostles in the *New Testament*, the Chinese *Book of Changes* (*I Ching*), the *Analects of Confucius* and the *Book of the Way* (*Tao Te Ching*) by Lao Tsu.

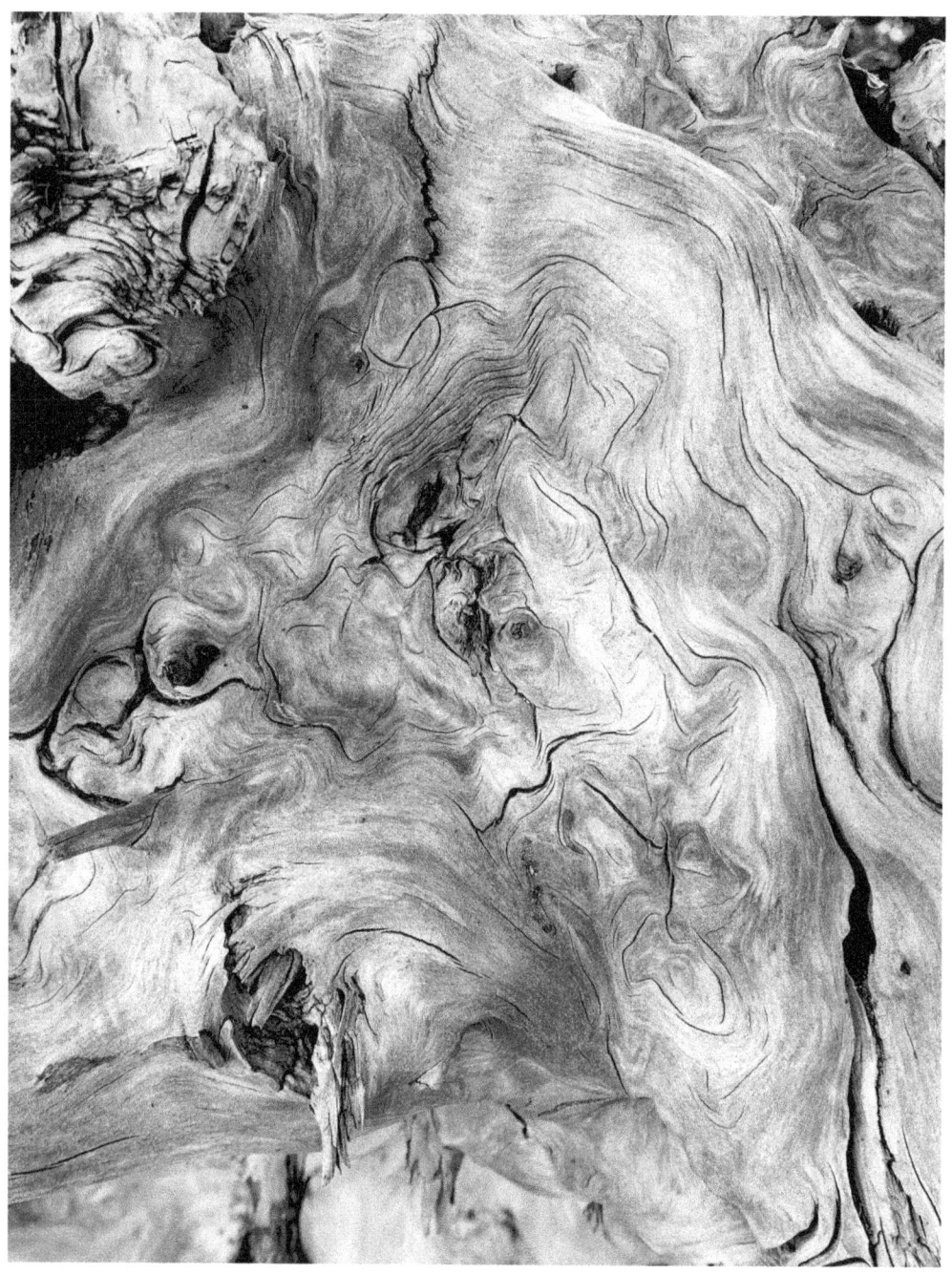

Bunnell, P. (2019). *Dynamic Temporal Architechture*. Photograph.

ASC
American Society for Cybernetics
a society for the art and
science of human understanding

Mutual Arisings:
Conversations with Humberto

Ray Ison[1]

In this guest ASC column I want to offer some reflections of a personal nature based on my encounters with Humberto Maturana in the years since we first met in 1988. It seems fitting to offer this reflection as a contribution to a special edition devoted to him. This is not the first time I have offered public reflections, and gratitude, for who I have become, and what I understand, due in large part to my conversations with Humberto (e.g., Ison, 2019). What follows is a kaleidoscope of moments of authentic conversation (Krippendorff, 2009), in a mutual arising of love in conversation—that is in the experience of being a legitimate other. The events are largely chronological. Please treat them as critical incidents (Flanagan, 1954), or significant moments in a 35 year punctuated conversation.

Figure 1. A hand drawing of Humberto Maturana done as part of an exercise at ASC 2001.

---

1. Email: ray.ison@open.ac.uk

Have you ever sat down opposite someone—on chairs facing each other—and attempted to draw what you see? In 2001 at the ASC conference in Vancouver I did this exercise with my partner, Humberto. The sketch (Figure 1) is what I produced, with the realization that people tend to forget that they also see their own lap and their drawing in progress. No doubt the designers of the exercise had particular motives in asking us to participate, though if I knew the reasons I am afraid I have forgotten. I can however say that I was left with a very precious memento signed by Humberto which now sits, framed on my Melbourne desk. It is a reminder of how much we see of others, and how little of ourselves, in intimate interactions if we are genuinely open to the other. I arise in my relations with others. If you wear glasses, as I do, then as the picture conveys one can be triggered to reflect on the means of one's own seeing. These considerations don't inform how we humans understand ourselves, especially through the lens of the totally misguided framing of humans as *Homo economicus* or as COVID-denying or vaccine-rejecting humans who mistakenly believe it is possible to be "free."

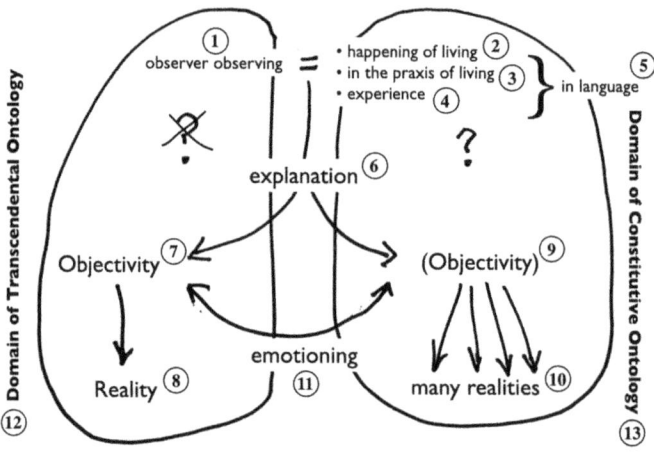

Figure 2. The design of one of Maturana's classic workshops. The numbers indicate the order of development of the ideas. (drawn by Pille Bunnell, based on a Maturana Workshop).

Humberto and I first met in Melbourne at one of his classic workshops where his presentation largely followed the schema in Figure 2. It proved a profound experience which heavily influenced the research I undertook with David Russell during the period 1989-2000 (see Ison & Russell, 2000; Russell & Ison, 2004, 2005). Humberto's development of a constitutive ontology continues to resonate in my current teaching and research.

In April 2022 I and some close colleagues will run the 10th of our Systems Thinking in Postgraduate Research systemic co-inquiries (Blackmore, Sriskandarajah,

& Ison, 2018), this time in conjunction with the 14th European Farming System Research conference in Evora, Portugal. Humberto's classic workshop sits at the core of our program design as it conjures an invitation to move from the pathway of objectivity to a pathway of responsibility in the doing of research. Rather than observers observing (Figure 2: number 1) the participants are invited to become researchers researching in an active, embodied way and thus to reflect on where they see themselves in their research situation: within, outside, or straddling a boundary. The possible pathways to that moment of a researcher arising as a researcher are then explored to reveal the pathways of objectivity or the pathways of responsibility (objectivity in parenthesis) and the choices one has as a researcher.

In 1997 Humberto accepted my invitation to visit the Systems Group at the Open University (UK). We have a filmed interview made during his visit; these YouTube segments come from an edited conversation between Humberto Maturana, myself and Raul Espejo (see Raul's paper pp. 63–76, this issue). The original filming was probably on Wednesday 12th March 1997. Humberto then presented a three-day workshop at Walton Hall from the 13-15th March. His visit to the OU is described in a Press Release (see below). I commend the first interview segment (Humberto Maturana's understanding and use of the term *system*) to those hung-up, or unclear about the ontological status of the concept system.

Press Release

March 1997 06

Humberto Maturana

One of the world's most eminent and radical biologists, epistemologist and systems thinker, is in the UK in the week beginning March 10 for an interdisciplinary event hosted by the Systems Department at the Open University. Maturana, Professor of Neuro-biology at the University of Chile is the central figure in what Fritjof Capra describes as the "Santiago School of Cognition." Maturana's work challenges many of the everyday things that we tend to take for granted but which, it could be argued, are at the heart of some of our contemporary ills. These include:

(i) **Objectivity** – Maturana's studies of cognition and his concern with the role of the observer, observing challenge notions of objectivity as currently held up by the scientific community. Maturana says "objectivity is a means to avoid responsibility."

(ii) **Information** – we live amongst pervasive metaphors which suggest we are in the "information revolution or age" and that better communication is about "information transfer." From a cognitive perspective these are unhelpful—information is derived from the latin *in formare* or "formed within." Maturana's research reveals just this—that there is no information from an external world but through our cognitive processes we impose patterns on the stimulatory signals we receive from the environment. Thus the environment triggers but does not determine what we call information. These insights pose major challenges to professions ranging from advertising to family therapy and teaching.

(iii) **Emotioning** – Maturana argues that our cognition derives from a braiding of our languaging and emotioning. Emotioning has a physiological basis. If emotioning is central to our cognition then this challenges the notions, common in science, that emotions are outside its domain of practice.

(iv) **Evolution** – Maturana's book *The Tree of Knowledge* (with Francisco Varela) challenged the dominant paradigm of evolution when it was first published in 1988.

Maturana's visit to the UK is a rare opportunity to discuss these themes at a time when they are subject to renewed focus following publication of Capra's new book—*The Web of Life*. Maturana is also recognised internationally as a radical epistemologist. His work has major implications for the ethics of action. As such Maturana's work has offered thinking that has the power to change the way we interact with the world.

Humberto was again a guest of the Open University and the SPMC (Systems Practice for Managing Complexity) Network in 2004. At St Anne's College Oxford on 6th September he led a one day workshop entitled "Fundamentalism, Ethics and Leadership" and on the same evening participated in a facilitated public conversation called "Biology, Love and Social Responsibility" with the OU's professor of biology, Stephen Rose. As the on-stage facilitator my memory is of a conversation that did not gel. There were several reasons for this, one of the main one's being Rose's lack of appreciation of Humberto's work beyond the concept autopoiesis. Recordings from this event are currently being digitised and will be available on the Systems Thinking in Practice Hub on the OU's OpenLearn platform[2] where all resources are free.

My meaningful memory from his 2004 visit was the conversation we had as I drove him from Heathrow to Oxford. Among diverse topics we agreed to agree on an explanation that God was a human invention designed to avoid responsibility.

A year later Humberto and I met in Quito, Ecuador where we had both been invited to give keynotes at the First International Congress in Systems Science & Technology (Systems Approach and Systems Research in Knowledge Management). In my talk "Epistemic Change in a Learning System: Curriculum Design for Managing Complexity" I spoke about the doing of systems thinking in practice as an activity very much like juggling. This notion became a metaphor that continues to guide our teaching praxis. After my talk Humberto introduced me to his concept of isophor which, he said, best described my juggler because a metaphor was a displacement of a concept from one domain to another rather than an evocation of the same dynamic in two different domains (e.g., juggling and doing systems). It took me some time to warm to his suggestion of isophor so that for my 2010 book (Ison, 2010) I sat on the fence but by the second edition I more assertively claimed that my use of the juggler was an isophor rather than a metaphor (Ison, 2017).

---

2. https://www.open.edu/openlearn/systems-thinking-hub

Figure 3. Maturana and his colleague from Matriztica, Ximena Dávila, in Sardinia, 2011. They are flanked by the author and Pille Bunnell.

We then met in Sardinia in 2011 (Figure 3) and California in 2012; after that, time, distance and opportunity intervened and kept our conversation muted. A conversation about theory in relation to practice, and thus praxis, remained in the mode of "to be continued" but it never was. There was much that we might have conversed about, had the moment arisen. Now I am left with my own responsibility, captured in Figure 4.

We live a systemic failure of responsibility: for what we do when we do what we do, and how we frame being human

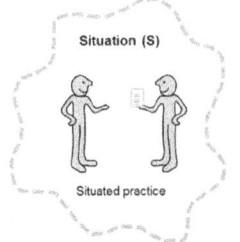

ii. all observations require a particular observer;
iii. everything said is said by someone to someone (we live in language, in its broadest sense);
iv. all knowing is doing;
v. all being, knowing & doing is relational (all is relational);
vi. all observers, practitioners, actors have a history, <u>a tradition of understanding,</u> from which they think & act
vii. institutions & technologies mediate practice
viii. we humans live with a desire for explanation - science is a practice which realizes scientific explanations
ix. a human with freedom is a social myth based on inadequate framing choices e.g. *Homo economicus*

Figure 4: Ison's key elements of responsibility, a basis of a praxeology for being a systems thinking practitioner, based on his Maturanan understandings

References

Blackmore, C., Sriskandarajah, N. & Ison, R. L. (2018). Developing learning systems for addressing uncertainty in farming, food and environment: What has changed in recent times? 13th European International Farming Systems Association (IFSA) Symposium, Greece. *International Journal of Agricultural Extension. 2018. 03-15*. Retrieved March 18, 2022 from https://journals.esciencepress.net/index.php/IJAE/article/view/2675/1312

Flanagan, J. (1954). The critical incident technique. *Psychological Bulletin, 51*(4), 327–358.

Ison, R. L. & Russell, D. B. (Eds.). (2000). *Agricultural extension and rural development: Breaking out of traditions*. Cambridge University Press.

Ison, R. L. (2019). Towards cyber-systemic thinking in practice. World *Futures, 75*(1/2), 5–16.

Ison, R. L. (2010). *Systems practice: How to act in a climate-change world*. Springer and The Open University.

Ison, R. L. (2017). *Systems Practice: How to Act in situations of uncertainty and complexity in a climate-change world* (2nd ed.). Springer and The Open University.

Krippendorff, K. (2009). Conversation. Possibilities for its repair and descent into discourse and computation. *Contructivist Foundations 4*(3), 138–150.

Russell, D. B., & Ison, R. L. (2005). The researcher of human systems is both choreographer and chorographer. *Systems Research and Behavioural Science 22*, 131–138.

Russell, D. B. & Ison, R. L. (2004). Maturana's intellectual contribution as a choreography of conversation and action. *Cybernetics & Human Knowing, 11*(2), 36–48.

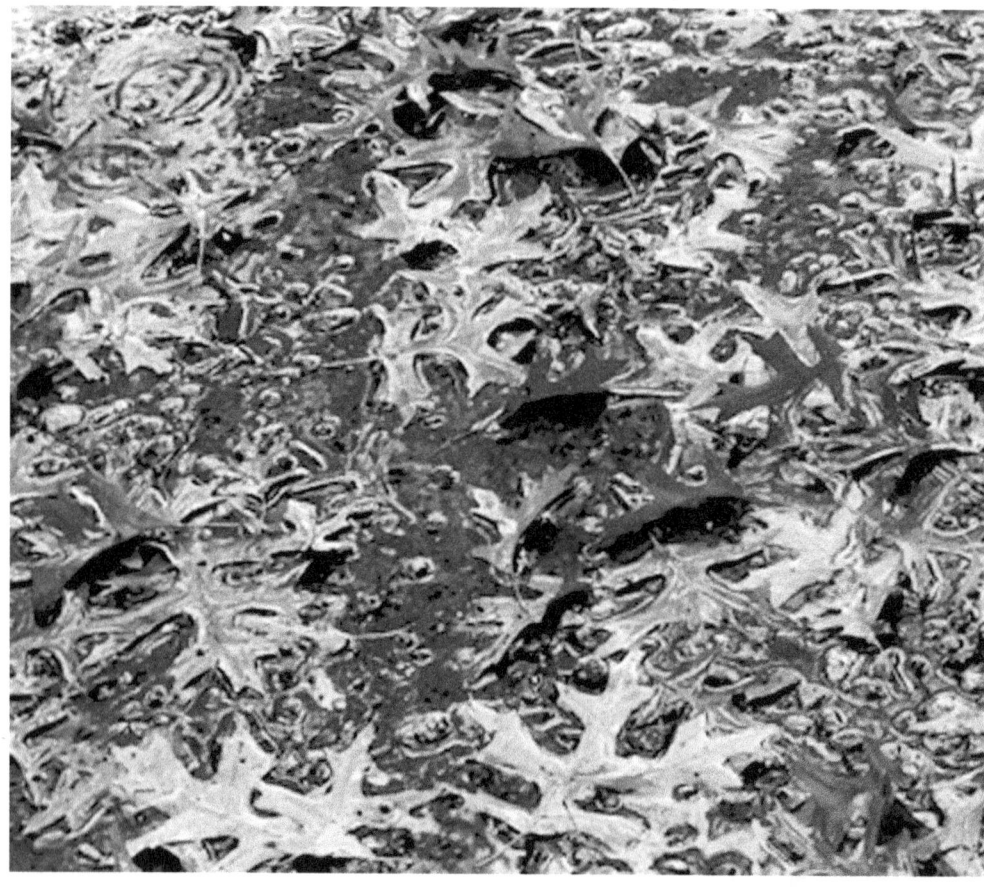

Bunnell, P. (2021). *Recomposition*. Photograph.

Leydesdorff's Compass

Mark William Johnson[1]

A Review of *The Evolutionary Dynamics of Discursive Knowledge* by Loet Leydesdorff. Series: Qualitative and Quantitative Analysis of Scientific and Scholarly Communication (W. Glänzel & A. Schubert, Eds.). 247 pp. Springer Nature (open-access). https://doi.org/10.1007/978-3-030-59951-5

Over the last 40 years, Loet Leydesdorff's project has been to develop practical and empirical applications of second-order cybernetic thinking with the aim of providing better metrics of the effectiveness of socio-economic organization. This is a project which has necessitated a deep reading not just of key cybernetic thinkers, but the philosophical and sociological provenance of cybernetic thinking, as well as technical development and application of models and algorithms. In recent history, there are few comparable examples of the depth and scope of this kind of intellectual enterprise—perhaps only those of some of the scholars to whom Leydesdorff's thinking is closest—most notably, Niklas Luhmann.

His new open-access book, *The Evolutionary Dynamics of Discursive Knowledge* is an important summation of the synthesis of Leydesdorff's ideas and practical investigative techniques. To talk of Leydesdorff's "compass" is to reflect two important aspects of this work: firstly, the richness and breadth of the intellectual territory he covers; and secondly, the coherence of his narrative and interpretation of his intellectual resources which enables him to navigate the reader, and here Leydesdorff has provided considerable clarity in the presentation of his ideas. Close reading rewards with profound insights into intellectual history and provides illumination on current trends such as *big data*, and popular avenues in contemporary social theory. The book's significance lies not just in the unique way Leydesdorff has reassembled and operationalized cybernetic sociology with Shannon's information theory, or how he has reunited cybernetics with its philosophical predecessors (particularly Husserl), but also the fact that Leydesdorff himself is a witness to the principal intellectual currents of late third of the 20th century and the opening decades of the 21st. While Luhmann was a personal friend, Leydesdorff's broader intellectual circle includes Latour, Habermas, Giddens and Simon as well other luminaries from cybernetics and evolutionary economics. Key intellectual moments are recounted—not least, the Habermas-Luhmann debate which has been so crucial to the future directions of sociology. This is against the backdrop of the social unrest of 1968, the Prague Spring (both of which he narrowly avoided getting mixed-up in!) through to the development of science and technology policies in the OECD and European Union.

---

1. Email: mj@ind.ku.dk

The *Evolutionary Dynamics of Discursive Knowledge* begins with an overview of themes which recur throughout the book. It sheds light on the social and philosophical context from which Leydesdorff's ideas come. For future generations of readers, this telling of the story of how ideas come to be thought is particularly important, and it makes for a compelling introduction to what follows. We are gently introduced to the back-story of Leydesdorff's thinking: from the challenges to Marxist dialectic presented by the information-driven economy, to 20th century phenomenology, and the fall-out from the Husserlian intersubjective view of consciousness, through to Parsons, Schutz, Luhmann and Simon. Behind all of it is the concept of meaning. Here we find one of the key themes in Leydesdorff: Husserl's *horizon of meaning* becomes a *horizon of expectations*, and with this, Leydesdorff is able to invoke the biology of anticipatory systems from Robert Rosen (1985), and a more mathematical representation produced by one of Rosen's students, Daniel Dubois. These fundamental elements are then gradually unpacked through the three parts of the book.

Part 1 deals with the concept of scientific knowledge and the way in which an empirics of knowledge can be operationalized. Sitting on Luhmann's shoulders allows Leydesdorff to present knowledge as a mechanistic communicative process which is grounded in selections of utterances steered by codes of communication which in turn produce the functionally differentiated society in which we all live. Scientific knowledge is one result of this functional differentiation, where what matters is what is selected as meaningful by groups of scientists. Leydesdorff convincingly argues that this selection of meaning leaves an imprint in the citation practices of scholars in academic journals. He argues that the modern sciences are discursive and mediated producing variation in texts and selection in citation practices. Citation for Leydesdorff provides a clear example of a process of codification—so citation practices are used as a vehicle for understanding the underlying dynamics of codes of communication in scholarly discourse.

This idea forms the basis of a large amount of empirical work which Leydesdorff has conducted using bibliometric data and other forms of text analysis. Following Luhmann, he argues that the sciences "self-organize into disciplines and specialties using specific codes" [p. 43], and the selection of meaning is determined by the operation of these codes. Part 1 explains some of the analytical techniques which can be used to expose a dynamics of meaning selection in science. Using Shannon's information theory (which ultimately is all about selection) Leydesdorff shows how Shannon's measure of transmission (otherwise called *mutual information*) in citation practices can act as an index of extent to which options are realized and organized within a system, while the unrealized options are represented by the generation of multiple versions of the same thing, which in Shannon is called *redundancy*. He argues that the complex dynamics of economic systems arise with the interaction between transmission and organization on the one hand, and the generation of redundancy and new options on the other. The dialectical tension between these underpins a major theme in the rest of the book.

Leydesdorff presents his thinking as dualist and not monist. In chapter 3 he criticizes Latour's actor-network theory because "heterogeneous dimensions are homogenized in a pan-semiosis" [p. 52]. Leydesdorff, by contrast, offers a structural, dualistic and dialectical approach based on different orders of system dynamics in communication: specifically, the dialectical relation between communication and expectation. He argues that, unlike in Latour (and indeed, a lot of postmodern theory, and some interpretations of second-order cybernetics), authors, texts and cognitions cannot be reduced to one another. Homogenizing dimensions leads in the direction of theology and the postulation of the single unifying cause for social complexity, or a prime mover. Leydesdorff argues that this cannot be right. While this assertion of irreducibility might lead one to suspect Leydesdorff of realism of the kind that Margaret Archer or Roy Bhaskar subscribe to, Leydesdorff remains true to the second-order cybernetic roots which he inherited from Luhmann. There is no mind-independent reality, but there are dialectical dynamics of communication and expectation from which consciousness constructs reality.

If Leydesdorff is to defend this position, however, he requires an empirics of expectation which can be distinguished from an empirics of communication. His transformation of Husserl's horizon of meanings into a horizon of expectations allows him to operationalize expectations in terms of anticipations. He argues that dualism between expectation and communication was always implicit in Husserl's *Cartesian Meditations*, but it was misread even by Husserl's most faithful followers (particularly Alfred Schutz). If there is a fundamental critique of modern social theory, it is here. Not just Latour is accused of monism: Anthony Giddens's structuration theory, which Leydesdorff builds on, suffers from monism for reasons similar to those indicated by Margaret Archer, who complains that Giddens is *elisionist* because he argues that the individual and the social are ontologically inseparable. (Archer, along with other Critical Realists disagree.) Giddens argued that structures could not be empirically investigated, because they represented "absent differences" (Giddens, 1979, p. 64). Leydesdorff argues to the contrary—the *second contingency* of expectation dynamics is empirically investigable: He asks, "Why would one not be allowed to formulate hypotheses about a second contingency in social structures?" [p. 57].

In the chapter "Towards a Calculus of Redundancy" [chapter 4], Leydesdorff shows how this might be done, by introducing the information theoretical calculations which have been at the centre of his empirical work. In revisiting Shannon's sender-receiver model, and challenging Shannon's own antipathy towards using his entropy calculations to measure meaning (Shannon's co-author, Warren Weaver, famously disagreed with this!), Leydesdorff elaborates higher-order dynamics in human communication wherein the selection mechanisms for meaning are constructed at a supra-individual level. He points out that at the center of his basic idea—that human communication depends on a deep level coordination of expectations, which operate on the basis of codes of communication and which have a self-organizing dynamic—what matters is the balance between the capacity to produce variation, or alternative options, and the process of selecting between options. The production of variation he sees as the generation of redundancy, arguing that this is particularly relevant to economic systems because, as he puts it: "The number of options available to an innovation system may be more decisive for its survival than the historically already-realized innovations" [p.27]. This therefore necessitates an indicator of surplus innovations, for which Leydesdorff deploys Shannon's understanding of redundancy. The dynamics between redundancies in scientific communication Leydesdorff relates to synergy.

The mechanics and exploitation of an indicator of synergy and surplus innovations is the focus for part 2 of the book. At the heart of Leydesdorff's empirical work has been economic analysis which explores the dynamics between different institutional groups, particularly on the triple helix relations between universities, industry and government. Perhaps the most obvious question to ask with regard to this is, why three? Leydesdorff responds (citing Simmel, 1902) that only with three dimensions do complex dynamics emerge. Each dimension establishes its identity through specific codes of communication, and innovation can be seen to be the result when different codes of communication interact or interfere with one another, producing varying levels of alternative descriptions of the same thing (mutual redundancy), and transmission (mutual information). This allows Leydesdorff to decompose the Shannon formulae for mutual information in three dimensions (which has puzzled cyberneticians since Ashby), identifying two complementary components: a measure of transmission, and a measure of mutual redundancy.

Part 2 gives an overview of the empirical application of these techniques, which have been applied to various national and regional economic systems. Throughout this work, Leydesdorff looks for synergies in the activities of researchers or innovators, looking specifically at the dynamics across regional boundaries, using a variety of data sources including the three levels of Eurostat Nomenclature of Territorial Units (NUTS) data. The mutual redundancy measure, derived from Shannon formulae, provides the indicator of the generation of multiple options between different organizations. This can be calculated at a basic level by examining co-authorship relations and using the entropies of citation statistics in the Shannon equations against geographical locations of research activities. This highlights certain geographical features which seem to be significant in driving synergies: for example, the highways to Amsterdam Airport, or development in the coastal regions of Norway driven by foreign investment. Chapter 6 presents a fully-worked example of these techniques examining the regional economic differences in Italy, which leads him to conclude that regional policies risk overlooking the significance of inter-regional development. This leads to a powerful political point:

> A political administration that is not reflexively aware of and informed about how the relevant innovation systems are shaped may lack the flexibility required to steer these systems and feel in the longer term constrained by the unintended consequences of its own actions. [p. 13]

More specific details of how Leydesdorff calculates the synergy measure are provided in chapter 7, where an illustrative toy model is presented. Fundamentally, this is a calculation of the entropy of terms in documents. So, for example, one might have a body of academic papers in which the occurrences of key terms are analysed. Across the whole dataset, levels of occurrence and co-occurrence can be calculated in terms of Shannon's entropy. With this, mutual information can be calculated between pairs of variables, while mutual redundancy can be calculated between three or more variables. If the latter calculation produces a number greater than the sum total of mutual information, this indicates a high level of synergy within the system. This presents ways in which degrees of synergy in co-authorship relations can be calculated across regional and national boundaries and comparisons made in relation to specific policy instruments in each case.

In part 3, Leydesdorff's focus turns from empirical work to simulation. To do this, he enriches the biological mechanism of autopoiesis of the social system (from Luhmann's theory), with the work on anticipatory systems by Robert Rosen and Daniel Dubois (1998). Anticipation is a biological dimension that is missing from Maturana's theory, but as others have argued (Conant & Ashby, 1970; Beer, 1995), the projection of future possible states is essential for the effective management of variety in a system. The domain of biological anticipation provides Leydesdorff with an alternative way of thinking about the dynamics of expectation in addition to his empirical analysis with Shannon equations. With the help of Daniel Dubois, who showed how an anticipatory system can be conceived and computed as a fractal, Leydesdorff shows how such a fractal can be simulated to demonstrate the operational principle of anticipation. It seems, if one is to predict future events on the basis of past events, there must be some way of determining a common pattern which connects past to future, and this common pattern must be fractal.

Like all fractals, the pattern which connects past to future is constructed as an interference pattern between different levels of dynamic process. In Dubois and Leydesdorff's scheme, there are three levels of dynamic: the lowest level concerns the flow of events in time, and the projection of future events on the basis of past events. Dubois calls this *recursion*. The second level concerns the reflexive construction of a model of the dynamics of events, which is called *incursion* (because the model feeds back on itself). The third level involves generation and selection of possible models of the future, which is called *hyperincursion*. Dubois's mathematics express this in terms of variations of the logistic equation, and drawing on joint work with Dubois, Leydesdorff shows how the interference between these levels can produce anticipatory behavior.

The simulations that Leydesdorff presents with these equations underlie the fundamental point he wishes to make about dualism between communication and horizons of expectation. The latter can be represented and simulated by the Dubois fractals, meaning that Leydesdorff can present a strong case to counter those like Giddens who argue for the impossibility to empirically investigate expectations, or Latour's homogenizing of dimensions. The horizon of expectations is the domain of unrealized possibilities which can be empirically identified through measuring redundancy, and simulated through the equations of Dubois. The dialectical dynamic between different processes can be made manifest and can be investigated. Furthermore, by focusing on the redundancy in the system, he is able to show how the dynamics of absence as causal force—which Deacon (2012) has drawn attention to—can be operationalized.

In one of the most profound sections towards the end of the book, Leydesdorff unpacks Descartes's intention behind "cogito ergo sum." Drawing on Heidegger's comment that *cogito ergo sum* was simply a statement of a first principle rather than any claim about the ontological status of thinking, Leydesdorff argues that the Cartesian dualism between mind and body can be reinterpreted as a dualism between first and second-order contingencies, between the selections of utterances and the codification of meaning. This is a powerful idea which reorients centuries of debate about mind and body: The essence of a Marxist dialectic can be maintained, without postulating a mind-independent reality, or what Bhaskar (1979) calls the *intransitive domain*.

Perhaps the most important intellectual attribute one must develop in an increasingly complex intellectual and technological environment is a good compass. Compasses are complex technological objects combining practical material properties, calculations and human expectations. Leydesdorff has made his compass out of a combination of practical empirical work, sophisticated simulation and a close reading of philosophy and sociology. What, for example, are we to make of the big data trend which threatens to swallow up social theory in a mass of algorithmic calculation? Leydesdorff's compass detects the problem which seems to escape the majority of discourse around data: It is monistic because behind the big data rationale is a proposition that ultimately everything is data. So again, there is reduction of social dynamics to a single causal agent. Where then is the dialectical pulse which drives social life? The same applies for those theories which reduce social complexity to biological causes, or to see society as a kind of meta-biology, as Habermas accused Luhmann of doing (with some justification).

Making the book available for free download is very welcome. The printed book is also very nicely produced with good use of color diagrams and excellent indices and referencing. I can imagine that it will appeal to sociologists, cyberneticians and economists, and anyone with an interest in the relationship between universities, society and innovation. Taken in its entirety, or in causal reading, there is something new for everyone. There are many intriguing references and original perspectives on other literature which scholars will find extremely valuable. Taken as a whole, it is a magisterial synthesis, which while it makes demands of the reader, will be guaranteed to provide fresh insights and perspectives on current thinking in sociology, technology, economics and cybernetics. Leydesdorff has pursued his logic relentlessly, backed-up both with detailed readings of his intellectual domain, and practical examples of how to use his equations in a meaningful way. This is refreshing when so much of today's sociological critique resorts to a kind of intellectual posturing, or data analysis is insufficiently grounded in an ontology and epistemology. Leydesdorff has shown how a deeper, coherent and empirical approach is possible.

References:

Beer, S. (1995). *Brain of the firm*. Chichester, UK: Wiley.
Bhaskar, R. (1979). *The possibility of naturalism*. Humanities Press.
Conant, R. & Ashby, R. (1971). Every good regulator must be a model of a system. *International Journal of Systems Science, 1*(2), 89–97.
Deacon, T. (2012). *Incomplete nature*. New York: Norton.
Dubois, D. M. (1998). Computing anticipatory systems with incursion and hyperincursion. In D. M. Dubois (Ed.), *Computing anticipatory systems, CASYS-first international conference* (Vol. 437, pp. 3–29). Woodbury, NY: American Institute of Physics.
Giddens, A. (1979). *Central problems in social theory*. London: Macmillan.
Simmel, G. (1902). The number of members as determining the sociological form of the group. *International American Journal of Sociology, 8*(1), 1–46.
Rosen, R. (1985). *Anticipatory systems: Philosophical, mathematical and methodological foundations*. Oxford, UK: Pergamon Press.

Bunnell, P. (2021). *My Tall Friends*. Photograph.

Bunnell, P. (2004). *Maturana Thoughtful*. Photograph.

www.ingramcontent.com/pod-product-compliance
Lightning Source LLC
Chambersburg PA
CBHW081421230426
43668CB00016B/2309